T0329523

Interval Analysis

Interval Analysis

Application in the Optimal Control Problems

Navid Razmjooy

IEEE PRESS

WILEY

Published by John Wiley & Sons, Inc., Hoboken, New Jersey.
Published simultaneously in Canada.

For general information on our other products and services or for technical support, please contact our Customer Care Department within the United States at (800) 762-2974, outside the United States at (317) 572-3993 or fax (317) 572-4002.

Wiley also publishes its books in a variety of electronic formats. Some content that appears in print may not be available in electronic formats. For more information about Wiley products, visit our website at www.wiley.com.

Library of Congress Cataloging-in-Publication Data applied for:

Hardback ISBN: 9781394190973

Cover Design: Wiley
Cover Image: © fotograzia/Getty Images

Set in 9.5/12.5pt STIXTwoText by Straive, Pondicherry, India

Contents

About the Author

Dr. Navid Razmjooy is a distinguished adjunct professor at Department of Computer Science and Engineering, Division of Research and Innovation, Saveetha School of Engineering, SIMATS, India. His main areas of research are the Renewable Energies, Machine Vision, Soft Computing, Data Mining, Evolutionary Algorithms, Interval Analysis, and System Control. Navid Razmjooy studied his Ph.D. in the field of Electrical Engineering (Control and Automation) from Tafresh University, Iran (2018). He is a senior member of IEEE/USA and YRC in IAU/Iran. He has been ranked among the world's top 2% scientists in the world based on the Stanford University/Scopus database. He published more than 200 papers and 8 books in English and Persian in peer-reviewed journals and conferences and is now Editor and reviewer in several national and international journals and conferences which can be found in https://www.webofscience.com/wos/author/rid/D-4912-2012. More information can be found in: https://www.researchgate.net/profile/Navid_Razmjooy.

1

Preface and Overview

Chapter 1: Preface and Overview

Mathematical modeling forms the backbone of scientific and engineering disciplines, enabling researchers to understand and solve complex real-world problems. These models provide a simplified representation of intricate systems, facilitating analysis and the development of practical solutions. However, balancing model simplicity and accuracy has always been challenging. While overly simplified models may lack the necessary precision, highly complex models often lead to equally complex problem-solving processes.

In recent years, advancements in mathematical modeling have made it possible to address these challenges through interval analysis. Interval analysis is a powerful tool that considers uncertainties within mathematical models, providing a more realistic and accurate representation of real-world systems. By expressing variables as intervals rather than single values, it becomes possible to account for uncertainties and analyze how they affect the overall system behavior.

This book explores the application of interval analysis in solving problems with interval uncertainties. It seeks to bridge the gap between overly simplified and overly complex models by providing a robust and practical approach to addressing uncertainties. By employing interval analysis techniques, researchers and engineers can obtain more reliable results and gain deeper insights into the behavior of complex systems.

The book begins with an introduction to mathematical modeling and the challenges of simplifying and understanding complex systems. It highlights the compromises made to balance model simplicity and accuracy. The subsequent chapters delve into the fundamentals of interval analysis, presenting topics such as the algebra of interval sets, interval representations, interval functions, and techniques for solving linear systems with interval parameters.

Interval Analysis: Application in the Optimal Control Problems, First Edition. Navid Razmjooy.
© 2024 The Institute of Electrical and Electronics Engineers, Inc.
Published 2024 by John Wiley & Sons, Inc.

Building upon this foundation, the book explores stability and controllability analysis based on interval analysis. It discusses techniques for testing stability, including the Routh–Hurwitz stability test and interval stability tests using linear matrix inequalities. Moreover, the book addresses controllability and observability concepts, shedding light on the essential properties of dynamic systems.

The book also covers the application of interval analysis in optimal control problems. It presents indirect and direct methods for solving optimal control problems and examines how these methods can be used to analyze and solve problems affected by interval uncertainties. Quadratic optimal control problems with interval uncertainties are also discussed, along with practical simulations to demonstrate the implementation of these approaches. What follows is a brief explanation of the chapters of this book.

Chapter 2: Introduction

This chapter delves into the fundamental concepts underlying mathematical modeling and its importance in analyzing real-world issues. It discusses how assumptions are often employed to simplify mathematical models, providing an overview of the compromises between model simplicity and precision. Additionally, it explores recent developments that have improved the accuracy and efficiency of these models, leading to more robust solutions for applied problems. However, it also highlights the potential limitations that arise as a result.

Chapter 3: Literature Review

Building upon the foundation laid in Chapter 2, this chapter focuses on reviewing the existing literature surrounding control systems. It emphasizes adjusting and controlling internal state variables to enhance system performance. The chapter also provides insights into classical control systems and their aim to improve specific functional characteristics such as transient behavior, settling time, and overshoot. Furthermore, it explores advanced studies wherein the optimal behavior of a system is achieved through the minimization or maximization of performance indices.

Chapter 4: Introduction to Interval Analysis and Solving the Problems with Interval Uncertainties

Chapter 4 introduces interval analysis and its role in handling problems affected by interval uncertainties. The chapter covers various topics, including the algebra of interval sets, interval representations, and interval functions; solving linear systems with interval parameters, interval derivatives, and integrals; and determining minimal intervals. Moreover, it explores advanced techniques such as the centered inclusion method, the Interval Runge–Kutta Method (IRKM) for interval differential equations, and interval uncertainty analysis based on orthogonal functions.

Chapter 5: Stability and Controllability Based on Interval Analysis

This chapter focuses on stability and controllability analysis using interval analysis. It explains how interval stability and controllability provide valuable insights into system behavior. The chapter investigates stability tests, such as the Routh–Hurwitz stability test and the interval Routh–Hurwitz stability test (Kharitonov Method). It also delves into interval stability based on linear matrix inequalities (LMIs). Furthermore, the concepts of controllability and observability are explored, shedding light on the essential properties of dynamic systems.

Chapter 6: Optimal Control of the Systems with Interval Uncertainties

In Chapter 6, the book focuses on optimal control of systems affected by interval uncertainties. It presents both indirect and direct methods for solving optimal control problems. The chapter examines how these methods can be employed to analyze and solve problems through techniques such as Euler–Lagrange equations, the interval Runge–Kutta method, the Chebyshev inclusion method, and the piecewise interval Chebyshev method (PICM). Quadratic optimal control problems with interval uncertainties and the interval quadratic regulator are also addressed based on indirect methods. Application-oriented simulations are included to illustrate the practical implementation of these approaches.

Chapter 7: Conclusions

The final chapter of the book serves as a summary and conclusion. It highlights the key findings, contributions, and limitations discussed in the preceding chapters. Furthermore, it emphasizes the importance of interval analysis in handling problems affected by uncertainties. The chapter concludes by suggesting potential areas of future research and the potential applications of the methods presented throughout the book.

In conclusion, this book comprehensively explores interval analysis and its application in solving problems affected by interval uncertainties. By striking a balance between model simplicity and accuracy, interval analysis offers robust solutions for addressing real-world complexities. This book will be valuable for researchers, scientists, and engineers seeking effective problem-solving techniques in diverse fields.

2

Introduction

2.1 Background

Mathematical models are widely utilized across various scientific and engineering fields to analyze and address real-world issues. However, the accuracy of these models can be compromised due to the inherent simplifications made during their development. To simplify the analysis process, assumptions are often employed, leading to a decrease in model precision. Conversely, as the pursuit of greater accuracy intensifies, the resulting models become more complex, rendering problem-solving equally intricate. Striking the right balance between model simplicity and precision is therefore crucial. In recent years, there have been significant developments aimed at improving the accuracy and efficiency of mathematical models.

These advancements have sparked researchers' interest in exploring new methods and approaches to achieve more robust solutions for applied problems. However, as we delve deeper into the complexities of real-world systems, limitations arise that can hinder progress and even bring it to a standstill. One critical challenge in modeling complex systems is the presence of uncertainties. Real-world scenarios are rarely deterministic, and interval uncertainties play a significant role in affecting their behavior. Interval analysis provides a robust framework for addressing these uncertainties within mathematical models. By expressing variables as intervals rather than single values, interval analysis allows for a more realistic representation of system behavior, considering the potential range of values rather than relying on point estimates.

The relevance of interval analysis in solving problems with uncertainties must be considered. It offers a means to balance oversimplified models that overlook uncertainties and overly complex models that hinder problem-solving.

Interval Analysis: Application in the Optimal Control Problems, First Edition. Navid Razmjooy.
© 2024 The Institute of Electrical and Electronics Engineers, Inc.
Published 2024 by John Wiley & Sons, Inc.

By incorporating interval analysis techniques, researchers and engineers can obtain more accurate and reliable results, accounting for the inherent uncertainties in real-world systems. This book explores the concept of interval analysis and its application in solving problems affected by interval uncertainties. It aims to provide researchers, scientists, and engineers with a comprehensive understanding of interval analysis techniques and their practical implementation.

By employing interval analysis, practitioners can enhance the accuracy and reliability of their models and gain valuable insights into the behavior of complex systems. Through carefully examining interval analysis methods and their applications, this book aims to address the limitations imposed by oversimplification and excessive complexity, ultimately facilitating more effective problem-solving in diverse fields. These limitations include the following:

1) A series of mathematical models include parameters, most of which cannot be accurately calculated in practice. There are many sources of parametric uncertainty, two of the most important of which are [1] as follows:
   - Measurement error is one of the largest sources of indeterminacy in parameter estimation, resulting from measurement instruments and environmental factors.
   - Errors in parameter estimation can occur due to incorrect classification or estimation of parameters with low or unspecified sample sizes. Consequently, this category introduces uncertainty into the model's parameter discussion.
2) Uncertainties in a real model are not the same type; therefore, each one should be appropriately identified within its domain, and its range should be defined. The uncertainties can be divided into three different classes:
   - *Epistemic uncertainty*: Mostly, there are several reasons for disregarding or lacking sufficient information about the physical system, environment, or estimation of system parameters, etc. [2].
   - *Aleatory (random) uncertainty*: This uncertainty is caused by random processes that arise from the nature of the actual system or are influenced by the environment, such as noise and similar scenarios.
   - *Stochastic uncertainty*: Stochastic uncertainty considers the system's high sensitivity to the initial conditions [3].
   - Each model (even without uncertainty) has its specific solution that cannot be applied to another one simultaneously. Random uncertainties are often represented using probability density functions (PDFs), whereas epistemic uncertainties are frequently depicted through fuzzy, interval, and stochastic variables.

For this reason, appropriate methods related to each type of problem should be used to obtain a reliable response from a system considering these uncertainties.

2.2 Relationship Between Error and Uncertainty

In general, the difference between the measured value and the actual value of a system is the calculated error value [4]. Some error values are generated outside the calculation; for instance, there are cases where the inputs are not adequately measured, or even some of the measured information is lost. Additionally, when the system modeling is based on simplification, it may ignore a large part of the system parameters, and these omissions are also considered uncertainties [5].

Other sources of error arise from internal resources based on the discrete nature of digital computing due to constraints such as computational time, storage capacity, or program complexity. Compatibility constraints, such as floating-point representation and conversion, also contribute to these issues.

These problems make it nearly impossible to perfectly align the original system with an approximate model, often involving reducing, rounding, and simplifying.

Precisely measuring the error of a numerical algorithm is typically challenging. In the real world, determining the exact value of the output error for a numerical program is impossible [6]. Consequently, attempts to find an approximate error to solve problems often prove unsuccessful. However, due to the widespread application of approximate engineering approaches, they have become valuable tools in engineering. Over the years, researchers have proposed various methods to handle and address these errors in the presence of uncertainties.

Among the many methods used in this field, Fuzzy methods [7, 8], statistical methods [9], and interval methods [10] are particularly popular.

Each of the mentioned methods has its disadvantages. For example, a Fuzzy method can perform well if we possess complete knowledge of the system or if an expert with extensive information about the scheme can guide the learning process and account for all unknown uncertainties.

Statistical methods are helpful when statistical data, such as mean value and variance, are available. However, this information is not always accessible. On the other hand, the interval methodology employs techniques that only require upper and lower bounds to assess the system, which is typically the case in most modeling scenarios. Figure 2.1 illustrates the variables involved in statistical, Fuzzy, and interval methods.

As depicted in Figure 2.1, the probability variable is determined using the PDF, the Fuzzy variable is based on its membership functions, and the interval variable only requires lower and upper range values. To gain a deeper understanding of the significance of the interval methodology and its distinctions from other methods, we will briefly summarize definitions provided by experts.

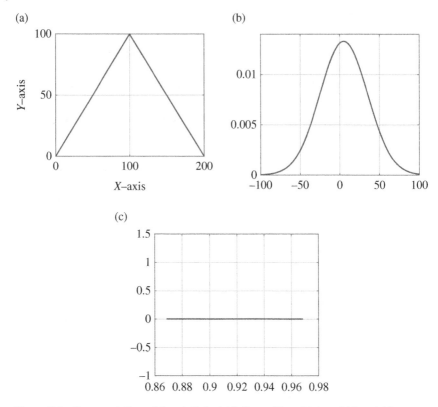

Figure 2.1 Representation of the statistical (a), Fuzzy (b), and interval (c) variables.

2.3 Expert Perspectives on Interval Analysis

- The difference between Fuzzy and interval sets is that a Fuzzy set has a membership function that estimates the extent (within the interval [0, 1]) to which any member belongs to the set. Interval sets, on the other hand, do not have such membership functions, and all the members of an interval are indistinguishable because we cannot say that any member holds a higher status in the set than the other.

- *This is true*: By definition, every algorithm that works for Fuzzy numbers can also be applied to interval numbers. The opposite is also true: thanks to Zadeh's extension principle; to compute, for example, the range Y of a Fuzzy unavailable $f(x_1, ..., x_n)$ over Fuzzy numbers $X_1, ..., X_n$, it is sufficient to compute, for each, the range $f(x_1(\alpha), ..., x_n(\alpha))$ over the crisp α-cuts $X_i(\alpha)$. This way, we obtain an alpha-cut $Y(\alpha)$ for Y. From this perspective, Fuzzy computations reduce to several instances of an interval computation problem. This is precisely how many books on Fuzzy (such as Klir and Yuan) introduce computing with Fuzzy numbers.

- From this viewpoint, it is not true that there are different algorithms for interval and Fuzzy variables. There are some papers in which the authors described algorithms for the general Fuzzy case, clearly understanding (and explicitly writing) that the interval case is a particular case. Some other researchers use the interval algorithms for alpha-cuts, thus getting Fuzzy algorithms.

- Contrary to the impression you seem to have, there are no separate interval and Fuzzy methods. We have two equivalent problems, and any form of solving one problem helps solve the other.

- From the research viewpoint, the problem of solving the system under interval (or Fuzzy) uncertainty is NP-hard, which means that no algorithm is perfect; you can constantly improve it. To show that what you are doing is good, you need to show that your algorithms are better than what was done before, and yes, you have to look both in the interval and Fuzzy literature since algorithms can be found in both places.

- There is no single problem of "solving interval system $Ax = b$." In contrast to the case of exactly known A and b, where the notion of a solution is clear, we have many different notions. For example, we have a united solution, the set of all possible solutions for all possible A and b; the controlled solution, the location of all x such that for all possible matrices A, we have Ax within the given ranges b, etc. There are similar different notions of Fuzzy solutions, which are more appropriate depending on the practical problem.

For example, Ohm's law says that $IR = V$ (where I is the current, R is the resistance, and V is the voltage). If you are measuring R, then from $I = [1, 2]$ and $V = [1, 3]$, you conclude only that R belongs to the united solution, which is in this case $[1/3, 2]$. If you are selecting a resistor for which, for all I from $[1, 2]$, the voltage should always be within $[1, 3]$, then the possible range of R is $[1,1.5]$, etc.

The second answer is that most methods of interval computations assume that the bounds are guaranteed, and exactly known. In the last example, that would mean that we have to be within the range $[1, 3]$, so if we take $I = 1.6$, this can potentially be a disaster, since we can get $2 \times 1.6 = 3.2$.

In Fuzzy theory, the endpoints of alpha-cut intervals are derived from experts and are not exact. As a result, many Fuzzy algorithms provide only an approximate solution, which is acceptable in the Fuzzy case, but may lead to mistakes when they are applied to intervals as special cases of Fuzzy sets.

2.4 Precision Analysis of Interval-Based Models (Self-Validated Numeric)

To solve a problem with uncertainty, numerous kinds of *Self-validated* methods have been proposed whereby implementing and simulating them, the computer follows the precision required in the program. Also, if the error size cannot be predicted in a response, it should be used with certain methods like interval analysis that encounter the error [11].

The most popular method among these methods is the interval method, which is introduced by Moore [12]. There are other methods, such as Hanson's generalized interval arithmetic [13] and Chernousko's Ellipsoid calculus [14]. The interval method, introduced by Moore is a technique where uncertain parameters in a model are represented by intervals with specified centers and radii. Ordinary methods are then extended to operate in the interval space. Instead of obtaining a single solution, a range of possible solutions is determined. This method is particularly suitable for bounded and pure sets.

Interval analysis enables the modeling of errors caused by physical measurements influenced by the environment. It can also characterize random uncertainties by representing their mean value and variance as intervals. By limiting the initial conditions within an interval and performing interval computations, the effects of stochastic uncertainties can be investigated and controlled.

In recent years, interval analysis has gained popularity due to its computational efficiency. Many analytical works have been developed to determine error boundaries and utilize interval Taylor models' polynomial expansions for static and specific dynamic problems.

Models with uncertainties utilized interval analysis as a very useful tool for the cursor of modeling errors due to physical measurements influenced by the environment.

In general, it can be said that the interval method is a suitable method for bounded and pure sets. For this reason, researchers are interested in using this approach for solving uncertain problems.

In the interval method, the uncertain parameters in the model are limited to a specified interval with a specified center and radius, and then the ordinary methods are expanded into the interval space.

In this case, instead of the solution, a range is defined that contains the solutions. Of course, it can be characterizing the random uncertainty and its mean value and the variance with interval representation. It is also possible to investigate and control the stochastic uncertainty by limiting the initial condition in an interval and analyzing its effects by the interval computations.

In the last decade, interval analysis has expanded rapidly due to its high computational efficiency [15]. Most of the established works have been analytically designed to determine the error boundary and the use of interval Taylor models' polynomial expansions to determine the response to static problems and specific dynamical problems. This technique can be used for optimal control applications. Optimal control involves mathematically optimizing a system to achieve the best performance within given constraints. It focuses on finding control inputs that minimize or maximize an objective function while satisfying system dynamics and constraints. Control theory analyzes and designs systems to attain desired behavior, employing various techniques for controlling dynamic systems. Uncertainties in optimal control stem from unknown factors, such as measurement errors, disturbances, and model inaccuracies. Interval analysis, a mathematical technique, handles uncertainties by representing variables as intervals. It permits rigorous analysis and propagation of uncertainties. Challenges arise from uncertainties, including suboptimal performance due to robustness issues, model inaccuracies, disturbances, noise, safety, and reliability concerns. To address these challenges, interval analysis can represent uncertainties, enable robust control design, facilitate uncertainty propagation, and assist verification and validation. By incorporating these approaches, interval analysis empowers the development of control strategies that effectively handle uncertainties, ensuring system stability, safety, and reliable performance.

2.5 Concepts of the Ordinary and Interval-based Optimal Control

Most control and optimal control theory techniques need to improve in solving problems with uncertainties in different forms of intervals, stochasticity, and chaos. Using analytical methods to solve these kinds of problems is very complicated or impossible. Using discrete or parametric methods also results in improper solutions with these uncertainties. Therefore, the need for methods that can solve the control problem (CP) and optimize dynamic systems is quite obvious.

In general, optimal control of a dynamic system is finding the best solution for optimizing (minimizing or maximizing) some indices that can be limited under different constraints so that the physical constraints of the system remain unchanged at the same time. In this case, the performance index, or in other words, the cost function, can be considered a desirable system characteristic.

Solving optimal control problems is generally complicated, especially with constraints and uncertainties. The problem of nonlinear optimal control (NOC), in contrast to linear optimal control (LOC), has no analytical methodology, which has led many researchers to find a solution to this problem. We can also accept the fact that a linear system is an estimate of a nonlinear model with uncertainty whose value is not always known. For this reason, obtaining an estimate for this type of system can be an inappropriate response for the real system. Therefore, the best way to optimal control the real-world system is to take into account the uncertainties of the system within a certain range that identifies the main goal of this study.

Over the past 40 years, the problem of optimal control has been solved by two approaches, including direct and indirect methods [16].

1) *Direct methods*: In these methods, the optimal solution is obtained by directly optimizing (minimizing or maximizing) the performance index based on the desired constraints. Applying these methods ultimately makes the main problem a linear or nonlinear optimization problem. Finally, the new problem is solved using linear programming (LP) or nonlinear programming (NLP) techniques. Direct methods are divided into two main groups:

   - *Discrete method*: It needs a large number of discrete points (samples) of the state variables and/or control variables to show accurate results which increases the size of the system.
   - *Parametric method*: The response can be based on expansion with basic functions such as polynomials or orthogonal functions [17].

This technique is performed with three parameterization techniques, including parameterization of the state, parameterization of the control, and parameterization of the control-state combination which, in any case, are estimated from a limited series of functions with unknown parameters.

2) *Indirect methods*: In these methods, the optimal control problem has been transformed into a boundary value problem (BVP) or initial value problem (IVP) with the Hamilton–Jacobi–Bellman (HJB) problem constraints [18]. In the long run, we expect the newly transformed problem to have a more straightforward solution than the original one.

   Problem conversion is often performed by using definitive methods. With the development of new issues in the field of interval uncertainty sources, we can propose a more robust method by modifying the existing methods based on interval analysis.

   The main advantage of the indirect methods is that the solution has unique and definite results; however, if the HJB problem is analytically solvable, an estimate of the error is numerically determinable [19]. In certain cases, where the performance index is quadratic and the linear equation is equal, the equation eventually becomes a system of differential equations (DEs) of Riccati, which is very popular in practice. Nevertheless, the indirect method has the following disadvantages [20]:

   - Solving NOC problems based on HJB is complicated.
   - The control system may not be resilient.
   - The designer should have a deep knowledge of the mathematics and physics of the system.

2.6   Orthogonal Spectral Methods

Over the past few years, spectral and pseudo-spectral methods have been widely used as one of the most popular parametric methods to obtain an approximate solution for solving problems that are associated with DEs and integral equations (IEs) and their combinations are used as CPs.

The basis of this method is the transformation of DEs and IEs into an algebraic equation [21]. In a CP, the state variable or control is expanded in terms of certain orthogonal functions, and then by derivative and integral generating matrices, the derivative and integral operations are eliminated from the original problem, and eventually converted to an algebraic equation.

Solving the algebraic equation obtains an estimate of the system response. The shape of the operational matrices depends on how the spectral functions (SFs) are chosen. Recently, orthogonal spectral functions (OSFs) have been widely used because of their efficiency and speed. The OSFs are divided into several categories [22]:

- *Piecewise orthogonal functions*: Walsh functions, pulse block functions, wavelet functions, etc.
- *Continuous orthogonal functions*: Fourier series, Bessel series transmitted Legendre, Chebyshev functions, Hermitian polynomials, etc.
- *Combined orthogonal functions*: A combination of continuous and piecewise fragmentation like Taylor and Block Pulse.

As mentioned before, using conventional methods for solving unknown systems will cause fundamental problems in their results, so we can use orthogonal methods to solve such problems.

As a result, an extension has to be established to the method of operation to the interval space. One of the other advantages of orthogonal functions is their ability to control overestimation. Therefore, combining this method with an interval method will show more compact and appropriate results for uncertain problems.

2.7   Desired Confidence Interval

At the beginning of the interval analysis introduction, the calculated bounds for solving the problems needed to be tight enough to use. However, further research by researchers has been done. In Figure 2.2, various methods for solving optimal control problems with uncertainties have been shown.

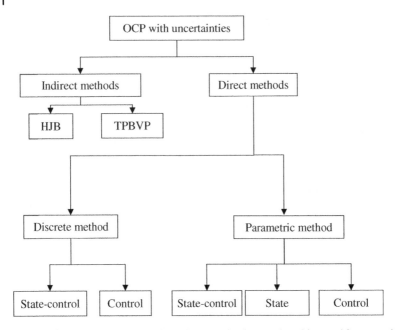

Figure 2.2 Different methods for solving optimal control problems with uncertainties.

One of the essential purposes of this book is to provide an improved parametric method for solving the optimal control problem with uncertainties and analyzing, designing, and implementing it.

2.8 Conclusions

In this chapter, we looked at the errors and intruders in generating uncertainties in physical models and introduced them. It was shown that uncertainties make the equation collapse, making the classical methods fail to solve these problems. Then, by introducing various methods for solving systems with uncertainties, we concluded that one of the best and easiest methods for analyzing these systems is the interval method.

In this chapter, useful explanations were provided on the benefits of the interval method over the other methods of analyzing uncertainties based on the correspondence with the experts in this field. In the following chapters, the spectral and orthogonal methods have been introduced, and the advantages of these methods for solving the derivative–integral equations are shown. Then, optimal control and its various forms have been introduced and how to consider the uncertainties in the optimal control problems has been shown.

References

1 Zitzenbacher, G., M. Längauer, and C. Holzer, *Modeling temperature and time dependence of the wetting of tool steel surfaces by polymer melts.* International Polymer Processing, 2017. **32**(2): p. 245–252.

2 Tang, Z.-C., M. J. Zuo, and Y. Xia, *Effect of truncated input parameter distribution on the integrity of safety instrumented systems under epistemic uncertainty.* IEEE Transactions on Reliability, 2017. **66**: p. 735.

3 Zeng, Z., et al., *A model-based reliability metric considering aleatory and epistemic uncertainty.* IEEE Access, 2017. **5**: p. 15505.

4 Strauch, B., *Investigating human error: incidents, accidents, and complex systems.* 2017: CRC Press.

5 Ebrahimi, N., et al., Tolerance analysis of the reflectarray antenna through Minkowski-based interval analysis, in *Antennas and Propagation (EUCAP), 2017 11th European Conference on*, 19–24 March 2017, Paris, France, IEEE. p. 2392–2395.

6 Waddington, C., *Should translations be assessed holistically or through error analysis?* HERMES-Journal of Language and Communication in Business, 2017. **14**(26): p. 15–37.

7 Peng, Z., J. Wang, and D. Wang, *Distributed maneuvering of autonomous surface vehicles based on neurodynamic optimization and fuzzy approximation.* IEEE Transactions on Control Systems Technology, 2017. **26**: p. 1083.

8 Yin, S., et al., *Adaptive fault-tolerant control for nonlinear system with unknown control directions based on fuzzy approximation.* IEEE Transactions on Systems, Man, and Cybernetics: Systems, 2017. **47**: p. 1909.

9 S. K. Jassim, and Z. I. AbdulNaby, *Statistical approximation operators.* Ibn AL-Haitham Journal For Pure and Applied Science, 2017. **26**(3): p. 350–356.

10 Zhang, C., et al., *A mixed interval power flow analysis under rectangular and polar coordinate system.* IEEE Transactions on Power Systems, 2017. **32**(2): p. 1422–1429.

11 Guerra, M. L., C. A. Magni, and L. Stefanini, *Average internal rate of return with interval arithmetic AMASES Proceedings*, UNIMORE, Italy. 2011.

12 Moore, R. E., R. B. Kearfott, and M. J. Cloud, *Introduction to interval analysis.* 2009: SIAM.

13 Hansen, E. R., A generalized interval arithmetic, in *Interval Mathematics.* 1975, Springer, p. 7–18.

14 Chernousko, F., *Ellipsoidal state estimation for dynamical systems.* Nonlinear Analysis: Theory, Methods & Applications, 2005. **63**(5): p. 872–879.

15 Viegas, C., et al., *Performance analysis and design of parallel kinematic machines using interval analysis.* Mechanism and Machine Theory, 2017. **115**: p. 218–236.

16 Cots, O., J. Gergaud, and D. Goubinat, *Direct and indirect methods in optimal control with state constraints and the climbing trajectory of an aircraft.* IRIT, Toulouse. 2016.

17 Li, B., and Y. Zhu, Parametric optimal control for uncertain linear quadratic models. Applied Soft Computing, 2017. **56**: p. 543–550.

18 Zhu, J., *A feedback optimal control by Hamilton-Jacobi-Bellman equation.* European Journal of Control, 2017. **37**: p. 70.

19 Nikoobin, A., and M. Moradi, *Indirect solution of optimal control problems with state variable inequality constraints: finite difference approximation.* Robotica, 2017. **35**(1): p. 50–72.

20 Pan, B., et al., *Double-homotopy method for solving optimal control problems.* Journal of Guidance, Control, and Dynamics, 2016. **39**: p. 1706.

21 Razmjooy, N., and M. Ramezani, *Analytical solution for optimal control by the second kind Chebyshev polynomials expansion.* Iranian Journal of Science and Technology, Transactions A: Science, 2017. **41**(4): p. 1017–1026.

22 Parand, K., M. Delkhosh, and M. Nikarya, *Novel orthogonal functions for solving differential equations of arbitrary order.* Tbilisi Mathematical Journal, 2017. **10**(1): p. 31–55.

3

Literature Review

In general, the variables of a physical system, known as state variables, change due to internal system changes and external control. Adjusting and applying proper management can enhance the performance of the system. Most of the attention in the study of classic control systems focused on improving some of the functional characteristics of the system, such as controlling the behavior of the transient state, reducing the settling time, and adjusting the amount of the overshoot, among other similar cases. In advanced studies, the system can be optimally desired by minimizing (maximizing) a performance index [1].

There are a lot of books and articles that overview solving optimal control problems [2]. One of these methods can be found in the proposed method by Richard Bellman, which was presented in 1950. This method was based on dynamic programming that provided the proper conditions for optimizing the Hamilton–Jacobi–Bellman equation.

In 1962, Pontrichen's progress for the maximal (minimal) principle provided a method for determining the optimal control of boundary problems known as the Bang-bang theory [3].

The aforementioned methods have been employed to solve complex optimal control problems, utilizing numerical techniques to tackle the intricacies involved in such problems. Recently, using spectral methods in analyzing more complex control systems has given special attention to researchers.

These methods offer two approaches to solving optimal control problems. The first approach is an indirect solution method, which involves converting the optimal control problem into a differential equation and solving it using spectral methods. The second approach is to provide a direct solution for the optimal control problem. This can be done by approximating the control variable, the state variable, or a combination of both variables. In general, there are numerous studies in both fields.

Interval Analysis: Application in the Optimal Control Problems, First Edition. Navid Razmjooy.
© 2024 The Institute of Electrical and Electronics Engineers, Inc.
Published 2024 by John Wiley & Sons, Inc.

For instance, in 1994, Dr. Razzaghi et al. [4] presented a method that utilizes orthogonal functions to solve time-variant linear quadratic optimal control problems. This reference was based on transforming the boundary-value double-point issues into a set of algebraic equations or transforming the optimal dynamic control problem into a quadratic programming problem.

In 1999, Park [5] in his thesis entitled "Chebyshev Estimates in Discrete and Spline Polynomials" investigated using the impulse method for calculating spline in discrete systems.

He explores how polynomials are produced using this method and its numerous applications in his dissertation.

The proposed method reduced the number of multiplications in the polynomial and increased its ability to apply numeric on the processor. According to Park's plan, Chebyshev's polynomials can produce the best estimation.

Because of the open point selection in the systems, there are several possible solutions, rather than an optimal response, which Rames' algorithm as a popular method cannot determine.

Through their examination of the proposed method, Park demonstrated that enhancing the spline using Chebyshev's method can lead to an optimal and distinct approach. In 2003, Pietz [6], in his dissertation entitled, "Pseudo-Spectral Coupling Methods for Direct Solving of Optimal Control Problems," introduced a method for discrete optimal control problems.

The aforementioned problems were considered generalized differential equations that turned into a nonlinear programming problem using the pseudo-spectral method. In his work, Legendre polynomials were used to achieve optimal control of the system. The final results were applied to both linear and nonlinear linear systems.

Jaddu and Vlach [7] used Chebyshev's polynomial and wavelet methods to solve constrained linear quadratic optimal control problems. This method was based on the transformation of the optimal control problem into mathematical programming, using a differential operating matrix for this purpose.

In 2007, Hennington [8] proposed another methodology for optimal control problems based on Gaussian parabolic spectral methods. According to the Hennington technique, orthogonal spectral methods have high ability and precision in optimal estimation and controlling of the optimal problems.

In this project, Hennington used the Gaussian method for control purposes in aerospace engineering. In Legendre, Rado and Gaussian were used for optimal control of the system. The primary purpose was to develop the Gaussian method.

In 2008, Tavallaei and Tousi [9] proposed a closed-loop methodology for solving linear optimal control systems using Legendre polynomials.

Garg [10] analyzed the pseudo-spectral method for optimal control of hybrid forms. Afterward, the characteristics of these two methods were investigated by studying the combination of Legendre–Gauss and Legendre–Gauss–Rado. The final results indicated that the stated methods could improve the optimum control of the system.

In 2013, Francolin et al. [11] proposed estimating the auxiliary mode for optimal control problems using the orthogonal arrangement in Gaussian square spots – the estimation of the additional state of optimal control problems using the Legendre–Gauss and Legendre–Gauss–Radu. The results indicate an improvement in the optimal control method.

In 2015, Kafash and Delavarkhalafi [12] introduced an algorithm based on state parameter parameterization for solving the optimal control problems on the Vanderpol oscillator.

The proposed method converted an optimal control problem to a solvable problem using time-dependent mathematical programming methods. The results indicated that the method was appropriate for optimal control purposes.

In 2017, Rafiei et al. [13] developed a method based on its combination with the wavelet method and achieved acceptable results. Analyzing and designing optimal control problems by optimizing a linear or quadratic performance index leads to the design of variable-state feedback. All physical processes have nonlinear behavior and various constraints.

Often, to simplify the systems in this case, the linear approximation of the model is used for conducting studies. Hence, determining the optimal control solution for these types of systems has accounted for most of the research.

As stated earlier, real (and optimal control) systems, even linear and quadratic, have some uncertainties in them that cannot provide a proper solution by deterministic methods. Usually, there were ways to solve this type of common problem.

Recently, some valuable methods have considered these uncertainties for solving physical problems [14]. For instance, the convex modeling method [15, 16], fuzzy method [17], intuitive theory [18], elliptic models [19, 20], and interval methods [21] are widely used to examine limited uncertainties.

Among these methods, the interval method stands out due to its inherent simplicity [22]. It is a convex model that defines the range of uncertainties based solely on identifying upper and lower limits [23]. Consequently, employing the interval method to establish a highly reasonable range of uncertainties appears justifiable.

In the interval methods, all possible values in an uncertain parameter are limited in a one-dimensional convex set that simplifies the measurement of an undefined parameter with no need for complete information. The definition of upper and lower bounds for an uncertain parameter is much easier than identifying a precision probability distribution for a random variable or membership functions in a fuzzy variable [24].

It should be noticed that most of the interval methods for solving static mechanical problems and structural systems include design optimization [25] and structural analysis problems [26, 27].

A limited number of studies have been presented on the application of interval analysis for solving ordinary differential equations (ODEs) or partial differential equations (PDEs), particularly with the performance index.

One of the methods for solving these problems is the interval Taylor series [28] as well as the inclusion Taylor model [29]. In 2017, Wu et al. [30] used a method based on chaos-based polynomials for a rigid and flexible part with uncertainty.

Until now, many sciences have been influenced by interval theory, and many successful scientific applications can be found in various fields, such as electrical engineering [31, 32], dynamic and turbulent systems [33], and control theory [34], among others. Rao and Berke [35] discussed uncertainty in engineering-design analysis problems and pointed out several cases in which an uncertain parameter can be expressed as an interval integer. In design and production, if a geometric parameter is accompanied by uncertainty, it can be modeled as an interval representation. Similarly, environmental parameters can be modeled as intervals if they are accompanied by uncertainty [36, 37].

Oppenheimer and Mitchel [38] provided a basis for analyzing the interval analysis of linear electric systems. Based on this idea, their first paper laid the foundations of interval analysis from an electrical engineering point of view [39].

Kolev [40], in his dissertation, reviewed interval analysis and its applications in circuit analysis [40]. Later, an interval method was used to analyze the error of electrical circuits in another paper [41].

In practical applications, due to the inevitability of inaccuracies in measurements, exact values of measured quantities often need to be taken into account. Instead, certain bounds are considered a possible interval. Therefore, interval matrices are used instead of ordinary matrices. Hansen and Smith were among the first researchers to use interval analysis in matrix calculations, and subsequent researchers continued their work in this field [42]. Interval matrices, their algebraic operations, their power, and their inverses can be used to describe the dynamics of systems, interval errors, time-independent fuzzy systems, and system analysis, as demonstrated by many references.

As is evident from the history of this work, it has been performed by two different groups of researchers: applied mathematics specialists and control engineering technicians. Both groups' efforts aimed at obtaining an optimal control system based on mathematical generalizations, with less uncertain control being taken into account.

Interval methods, like discrete analytical methods, do not solve problems with a unique algorithm that can be compared in speed or accuracy. The interval method is more likely to guarantee the correctness of the solution; therefore, the interval method is more known as a validation method for the interval containing the key.

To emphasize the importance of the problem, we can consider Warwick Tucker's research at the University of Cornell, which received several world awards, including the World Mathematical Society Prize in 2004 and the University of Michigan One-time Special Prize in the name of the World Moore Prize for outstanding researchers in the field of interval analysis.

In the following section, we will provide a brief overview of some specific cases in the control field in the presence of uncertainties.

In 2002, Jaulin et al. [43] proposed a methodology for identifying and controlling the system using the PI controller. In 2004, Ashokaraj et al. [44] used a method to estimate the state of a robot using a combination of improved Kalman filter method and open mathematics.

In 2010, Patre and Deore [45] reviewed a method based on state feedback for interval systems. To obtain the controller, the same general method was considered and used to extend the interval method on it in a way similar to the Kharitonov method.

In 2007, Ribot et al. [46] provided a method for solving nonlinear differential problems for state estimation. Their purpose was to consider system-induced disturbances as uncertainties and to estimate the state based on their method. Finally, they interpreted their method by applying the proposed method to a rocket model.

In 2012, Jaulin [47] proposed a new method for estimating the state of a nonlinear robot. The proposed method was a combination of the general method for estimating the state of nonlinear systems with the interval method. Eventually, they examined two systems related to the robotic sailboat system.

In 2015, Berkani et al. [48] presented a robust method based on interval calculations to control a nuclear reactor. They designed a PID controller and obtained acceptable results after modeling the reactor and considering a structural interval uncertainty.

In 2015, Leal et al. [49] focused on optimal control using the decision process and pursued the same goal as presented at the conference on computational mathematics and applications.

In complicated problems in the presence of uncertainties, the main goal is to carefully apply the interval optimal control over the open loop system. Among the applications of these types of systems, two commonly used washing machines, and traffic lights are the most applicants.

Because their existence may cause many changes in the results of the open loop optimal control problems and remove the system from the steady state or to its original purpose.

In recent decades, powerful methods have been introduced to consider uncertainties for optimal control problems with closed-loop strategy [50]. For example, a robust control method has been developed to reduce the system's sensitivity to uncertainties [51, 52].

However, the range of uncertain effects in an optimal open loop control system, especially in some designs, such as the tracking system's design, can be seen [53].

It is worth recalling that the methods introduced by feedback control, such as resistant control, do not relate to the domain of diffusion and diffusion of uncertainties in a system.

For instance, the domain of diffusion and uncertainty diffusion of the dynamic model, initial conditions, and dynamic model parameters are important research topics in the optimal control problem [54].

There are several methods for solving open-loop optimal control problems with uncertainties. Almost all methods proposed for these problems are probabilistic methods, such as stochastic optimal control, and polynomial chaos (PC)-based approximation methods.

Among probabilistic methods, stochastic optimal control methods have attracted much attention [55]. Here, the state variable in a controlled system over a while is assumed to be a random process and this random process cannot be measurable. In addition, the measurement itself is also usually subject to noise effects.

The most common method for solving optimal control problems with stochastic methods is the principle of Pontryagin's optimality and Bellman's dynamic programming. These two methods include two classes of the same, but they are equivalent.

The theoretical basis for the maximum random rule was proposed by Peng [56], which can be used to solve optimal control problems with random inputs.

Although significant advances have been made in random access control methods, all of them are based on the assumption that the distribution of the unknown input parameter is determined and the optimal goal is to minimize the final or integral costs.

In addition to this method, many probabilistic approximation methods were also proposed based on the Monte Carlo method. These methods were based on the principle that the distribution of uncertainty input parameters is determined.

The nature of these methods is that they use the statistical sampling method to model the problem, while the expected value of the function is approximated by various methods. Cao et al. [57] proposed a Monte Carlo method using the method of variance reduction and derivative of state-sensitive with uncertain parameters.

Kleywegt et al. [58] considered optimal control problems in the presence of uncertainties as a random discrete optimization problem, which was then solved by Monte Carlo's average sample approximation.

In 2014, Dorociak and Gausemeier [59] transformed the optimal control problem with uncertainties into a multi-objective optimization problem in which two objective functions were used based on a simple approximation to determine the Pareto boundary.

In 2016, Phelps et al. [60] provided more analysis and proof of compatibility results of the optimal control problem and the problem of the nonprecise control optimization, in which the computational framework was based on the average approximation of the samples using the Monte Carlo method.

PC approximation methods are commonly used indirect methods that perform better than Monte Carlo simulations in terms of computational cost.

Xiu and Karniadakis [61] presented a comprehensive chaotic polynomial structure (gPC) in which the Wiener–Askey orthogonal polynomial was used to illustrate various chaotic processes and the PC method was used to approximate different distributions of the problem [62].

Hover used the PC method to replace random and free variables in the path optimization problem [63].

Fisher and Bhattacharya [64] and Huschto and Sager [65] used the chaotic polynomial expansion method to convert the randomized optimal control problem to the determined optimal control problem. Then, this method was applied to the path planning problem in constrained mechanical systems and the Markov decision rule maintenance feature.

The central assumption for controlled models in all the stated methods is that they are Aleatory uncertainties.

However, in most optimal control problems, there may be uncertain or incomplete information. These uncertainties can be attributed to the various factors mentioned in the introduction. Although probabilistic methods are effective when complete statistical information is available about uncertain parameters, these methods are incapable when our only information is limited to the upper and lower bounds [66].

Many nonprobabilistic methods are available to solve these types of uncertainties, and, according to the explanations given before, the interval method is one of the best of them.

Direct use of interval analysis leads to an overestimation and often suffers from the high volume of computation [67]. To compensate for this shortcoming, an indirect method based on interval analysis can be used.

For example, the fuzzy-interval combinatorial method proposed by Dubois et al. [68] or an interval quadratic programming method in [69] is used for a particular category of interval models to minimize its parameters. However, the interval method is rarely used to control uncertainties in optimal control problems.

However, in 2017, Campos et al. did a related research. They proposed a discrete method for optimal control problems with interval uncertainties. In this study, they used a dynamic programming method to solve discrete-type optimal control problems; in this way, after developing the traditional dynamic programming to the interval dynamic programming, it was used to optimize the optimal control problems with uncertain parameters [70].

The boundary restriction function is a suitable tool for obtaining the result of the diffusion domain for an optimal control problem with interval uncertainties. An example is about using interval analysis for optimal control based on Hukuhara derivation. Leal et al. [71] take into account optimum control issues with an objective function with interval values. To acquire the essential conditions for this issue, we take into consideration two conceptions of order relation on the interval space. Using the generalized Hukuhara derivative (gH-derivative) notion for interval-valued functions, we can establish the essential requirements.

In this book, a range of analysis methods based on interval analysis have been employed to address the optimal control problems including uncertain distribution parameters. Then, a nonprobabilistic boundary method with the aid of Chebyshev functions will be presented to estimate the boundaries of compact spaces of state and control variables.

References

1 Kirk, D. E., *Optimal control theory: an introduction*. 2012: Courier Corporation.

2 Lin, Q., R. Loxton, and K. L. Teo, *The control parameterization method for nonlinear optimal control: a survey*. Journal of Industrial and Management Optimization, 2014. **10**(1): p. 275–309.

3 Kalman, R., *The theory of optimal control and the calculus of variations*. Mathematical Optimization Techniques, 1963. **309**: p. 309–331.

4 Razzaghi, M., and G. N. Elnagar, *Linear quadratic optimal control problems via shifted Legendre state parametrization*. International Journal of Systems Science, 1994. **25**(2): p. 393–399.

5 Park, J. H., *Chebyshev approximation of discrete polynomials and splines*, PhD dissertation. 1999: Virginia Polytechnic Institute and State University.

6 Pietz, J. A., *Pseudospectral collocation methods for the direct transcription of optimal control problems*. 2003: Rice University.

7 Jaddu, H., and M. Vlach, *Wavelets-based approach to optimizing linear systems*. Japan Advanced Institute of Science and Technology, Ishikawa, 2006: p. 923–1292.

8 Huntington, G. T., *Advancement and analysis of a Gauss pseudospectral transcription for optimal control problems*, Doctoral dissertation. **69**. 2007: Massachusetts Institute of Technology, Department of Aeronautics and Astronautics.

9 Tavallaei, M. A., and B. Tousi, *Closed form solution to an optimal control problem by orthogonal polynomial expansion*. American Journal of Engineering and Applied Sciences, 2008. **1**(2): p. 104–109.

10 Garg, D., *Advances in global pseudospectral methods for optimal control*. 2011: University of Florida.

11 Francolin, C. C., *Costate estimation for optimal control problems using orthogonal collocation at Gaussian quadrature points*. 2013: University of Florida.

12 Kafash, B. and A. Delavarkhalafi, *Restarted state parameterization method for optimal control problems*. Journal of Mathematics and Computer Science-JMCS, 2015. **14**(2): p. 151–161.

13 Rafiei, Z., B. Kafash, and S. Karbassi, *A new approach based on using Chebyshev wavelets for solving various optimal control problems*. Computational and Applied Mathematics, 2017: p. 1–14.

14 Moens, D., and D. Vandepitte, *A survey of non-probabilistic uncertainty treatment in finite element analysis*. Computer Methods in Applied Mechanics and Engineering, 2005. **194**(12): p. 1527–1555.

15 Jiang, C., X. Han, and G. Liu, *Optimization of structures with uncertain constraints based on convex model and satisfaction degree of interval*. Computer Methods in Applied Mechanics and Engineering, 2007. **196**(49): p. 4791–4800.

16 Kang, Z. and Y. Luo, *Non-probabilistic reliability-based topology optimization of geometrically nonlinear structures using convex models*. Computer Methods in Applied Mechanics and Engineering, 2009. **198**(41): p. 3228–3238.

17 de Barros, L. C., R. C. Bassanezi, and W. A. Lodwick, *Fuzzy sets theory and uncertainty in mathematical modeling*, in *A First Course in Fuzzy Logic, Fuzzy Dynamical Systems, and Biomathematics*. 2017, Springer. p. 1–21.

18 Shafer, G., *A mathematical theory of evidence*. Vol. 1. 1976: Princeton University Press Princeton.

19 Chernousko, F., *Ellipsoidal state estimation for dynamical systems*. Nonlinear Analysis: Theory, Methods & Applications, 2005. **63**(5): p. 872–879.

20 Chernousko, F., and D. Y. Rokityanskii, *Ellipsoidal bounds on reachable sets of dynamical systems with matrices subjected to uncertain perturbations1*. Journal of Optimization Theory and Applications, 2000. **104**(1): p. 1–19.

21 Alefeld, G. and G. Mayer, *Interval analysis: theory and applications*. Journal of Computational and Applied Mathematics, 2000. **121**(1): p. 421–464.

22 Corriou, J.-P., *Stability analysis*, in *Process Control*. 2018, Springer. p. 117–142.

23 Jiang, C., et al., *An uncertain structural optimization method based on nonlinear interval number programming and interval analysis method*. Engineering Structures, 2007. **29**(11): p. 3168–3177.

24 Revol, N., K. Makino, and M. Berz, *Taylor models and floating-point arithmetic: proof that arithmetic operations are validated in COSY*. The Journal of Logic and Algebraic Programming, 2005. **64**(1): p. 135–154.

25 Li, F., et al., *An uncertain multidisciplinary design optimization method using interval convex models*. Engineering Optimization, 2013. **45**(6): p. 697–718.

26 Chen, S., H. Lian, and X. Yang, *Interval static displacement analysis for structures with interval parameters.* International Journal for Numerical Methods in Engineering, 2002. **53**(2): p. 393–407.

27 Gao, W., *Natural frequency and mode shape analysis of structures with uncertainty.* Mechanical Systems and Signal Processing, 2007. **21**(1): p. 24–39.

28 Cheng, J., et al., *Direct reliability-based design optimization of uncertain structures with interval parameters.* Journal of Zhejiang University-SCIENCE A, 2016. **17**(11): p. 841–854.

29 Zheng, Z.J., et al. *Medium and long term load forecasting of power system based on interval taylor model arithmetic.* in *Advanced Materials Research.* 2014, Trans Tech Publ.

30 Wu, J., et al., *Uncertain dynamic analysis for rigid-flexible mechanisms with random geometry and material properties.* Mechanical Systems and Signal Processing, 2017. **85**: p. 487–511.

31 Manica, L., P. Rocca, and A. Massa, *Robust interval-based analysis of pattern distortions in reflector antennas.* in *Antennas and Propagation Society International Symposium (APSURSI), 6–11 July 2014, Memphis, TN, USA,* IEEE.

32 Pereira, L. and V. Da Costa, *Interval analysis applied to the maximum loading point of electric power systems considering load data uncertainties.* International Journal of Electrical Power & Energy Systems, 2014. **54**: p. 334–340.

33 Monti, A., F. Ponci, and M. Valtorta, *Extending polynomial chaos to include interval analysis.* IEEE Transactions on Instrumentation and Measurement, 2010. **59**(1): p. 48–55.

34 Giusti, A. and M. Althoff. *Efficient Computation of Interval-Arithmetic-Based Robust Controllers for Rigid Robots.* in *Robotic Computing (IRC), IEEE International Conference on 6–11 July 2014, Taichung, Taiwan,* IEEE.

35 Rao, S. S., and L. Berke, *Analysis of uncertain structural systems using interval analysis.* AIAA Journal, 1997. **35**(4): p. 727–735.

36 Shim, K.-S., et al., *Analysis of low frequency oscillation using the multi-interval parameter estimation method on a rolling blackout in the KEPCO system.* Energies, 2017. **10**(4): p. 484.

37 Simić, V., S. Dabić-Ostojić, and N. Bojović, *Interval-parameter semi-infinite programming model for used tire management and planning under uncertainty.* Computers & Industrial Engineering, 2017. **113**: 487.

38 Oppenheimer, E. P. and A. N. Michel, *Application of interval analysis techniques to linear systems. III. Initial value problems.* IEEE Transactions on Circuits and Systems, 1988. **35**(10): p. 1243–1256.

39 Leenaerts, D., *Application of interval analysis for circuit design.* IEEE Transactions on Circuits and Systems, 1990. **37**(6): p. 803–807.

40 Kolev, L., *Interval methods for circuit analysis.* Vol. 1. 1993: World Scientific.

41 Femia, N. and G. Spagnuolo, *True worst-case circuit tolerance analysis using genetic algorithms and affine arithmetic.* IEEE Transactions on Circuits and Systems I: Fundamental Theory and Applications, 2000. **47**(9): p. 1285–1296.

42 Yang, X., J. Cao, and J. Liang, *Exponential synchronization of memristive neural networks with delays: interval matrix method*. IEEE Transactions on Neural Networks and Learning Systems, 2017. **28**: 1878–1888.

43 Jaulin, L., I. Braems, and E. Walter. *Interval methods for nonlinear identification and robust control*. in *Decision and Control, 2002, Proceedings of the 41st IEEE Conference on* 10–13 December 2002, Las Vegas, NV, USA, IEEE.

44 Ashokaraj, I., et al., *Mobile robot localisation and navigation using multi-sensor fusion via interval analysis and UKF*. Proceedings of the 2004 Towards Autonomous Robotic Systems (TAROS), University of Essex, Colchester, UK, 2004.

45 Patre, B. M. and P. J. Deore, *Robust state feedback for interval systems: an interval analysis approach*. Reliable Computing, 2010. **14**(1): p. 46–60.

46 Ribot, P., C. Jauberthie, and L. Travé-Massuyes. *State estimation by interval analysis for a nonlinear differential aerospace model*. in *Control Conference (ECC), 2007 European*. 2–5 July 2007, Kos, Greece, IEEE.

47 Jaulin, L., *Combining interval analysis with flatness theory for state estimation of sailboat robots*. Mathematics in Computer Science, 2012: p. 1–13.

48 Berkani, S., F. Manseur, and A. Maidi, *Optimal control based on the variational iteration method*. Computers & Mathematics with Applications, 2012. **64**(4): p. 604–610.

49 Leal, U. A. S., G. N. Silva, and W. A. Lodwick, *Multi-objective optimization in optimal control problem with interval-valued objective function*. Proceeding Series of the Brazilian Society of Computational and Applied Mathematics, 2015. **3**(1): p. 010130-1–010130-7.

50 Dorf, R. C. and R. H. Bishop, *Modern control systems*. 2011: Pearson.

51 Dullerud, G. E. and F. Paganini, *A course in robust control theory: a convex approach*. Vol. 36. 2013: Springer Science & Business Media.

52 Parnianifard, A., et al., *An overview on robust design hybrid metamodeling: advanced methodology in process optimization under uncertainty*. International Journal of Industrial Engineering Computations, 2018. **9**(1): p. 1–32.

53 Loxton, R., K. L. Teo, and V. Rehbock, *Robust suboptimal control of nonlinear systems*. Applied Mathematics and Computation, 2011. **217**(14): p. 6566–6576.

54 Zazzera, F. B., et al., *Assessing the accuracy of interval arithmetic estimates in space flight mechanics*. Final Report, Ariadna id **4**(4105): p. 1–189.

55 Paulson, J. A. and A. Mesbah, *An efficient method for stochastic optimal control with joint chance constraints for nonlinear systems*. International Journal of Robust and Nonlinear Control, 2018. **29**: p. 5017.

56 Peng, S., *A general stochastic maximum principle for optimal control problems*. SIAM Journal on Control and Optimization, 1990. **28**(4): p. 966–979.

57 Cao, Y., M. Y. Hussaini, and T. A. Zang, *An efficient Monte Carlo method for optimal control problems with uncertainty*. Computational Optimization and Applications, 2003. **26**(3): p. 219–230.

58 Kleywegt, A. J., A. Shapiro, and T. Homem-de-Mello, *The sample average approximation method for stochastic discrete optimization*. SIAM Journal on Optimization, 2002. **12**(2): p. 479–502.

59 Dorociak, R. and J. Gausemeier, *Methods of improving the dependability of self-optimizing systems*, in *Dependability of Self-Optimizing Mechatronic Systems*. 2014, Springer. p. 37–171.

60 Phelps, C., J. O. Royset, and Q. Gong, *Optimal control of uncertain systems using sample average approximations*. SIAM Journal on Control and Optimization, 2016. **54**(1): p. 1–29.

61 Xiu, D. and G. E. Karniadakis, *The Wiener – Askey polynomial chaos for stochastic differential equations*. SIAM Journal on Scientific Computing, 2002. **24**(2): p. 619–644.

62 Askey, R. and J. A. Wilson, *Some basic hypergeometric orthogonal polynomials that generalize Jacobi polynomials*. Vol. 319. 1985: American Mathematical Soc.

63 Hover, F. S., *Gradient dynamic optimization with Legendre chaos*. Automatica, 2008. **44**(1): p. 135–140.

64 Fisher, J. and R. Bhattacharya, *Optimal trajectory generation with probabilistic system uncertainty using polynomial chaos*. Journal of Dynamic Systems, Measurement, and Control, 2011. **133**(1): p. 014501.

65 Huschto, T. and S. Sager. *Stochastic optimal control in the perspective of the Wiener chaos*. in *Control Conference (ECC), 2013 European*. 17–19 July 2013, Zurich, Switzerland, IEEE.

66 Wu, J., et al., *Interval uncertain method for multibody mechanical systems using Chebyshev inclusion functions*. International Journal for Numerical Methods in Engineering, 2013. **95**(7): p. 608–630.

67 Ishizaki, T., et al., *Interval quadratic programming for day-ahead dispatch of uncertain predicted demand*. Automatica, 2016. **64**: p. 163–173.

68 Dubois, D., et al., *Fuzzy interval analysis*, in *Fundamentals of fuzzy sets*. 2000, Springer. p. 483–581.

69 Liu, S.-T. and R.-T. Wang, *A numerical solution method to interval quadratic programming*. Applied Mathematics and Computation, 2007. **189**(2): p. 1274–1281.

70 Burghes, D. N. and A. Graham, *Introduction to control theory, including optimal control*. 1980: John Wiley & Sons.

71 Leal, U. A. S., G. N. Silva, and W. A. Lodwick, Necessary condition for optimal control problem with interval-valued objective function. Proceeding Series of the Brazilian Society of Computational and Applied Mathematics, 2015. **3**(1): 010131-1–010131-7.

4

Introduction to Interval Analysis and Solving the Problems with Interval Uncertainties

4.1 Introduction

The analysis of mathematical problems can be classified into three perspectives [1–3]:

1) *Continuous (infinite) arithmetic*: If the analysis of problems is based on the real set numbers $\mathbb{R}$ in the ideal world. The purpose of continuous arithmetic is to prove the existence and uniqueness of the system's solution and its analytic solution. Some examples of this arithmetic include the following: solving a system of linear equations using matrix operations in linear algebra, finding the maximum and minimum values of a function using calculus, and proving the convergence of a sequence in real analysis.

2) *Discrete (finite) arithmetic*: This kind of arithmetic identifies the problems in the machine space by discretizing them which is based on real discrete finite discrete spaces with a set of machine numbers M. The purpose of this type of calculation is to obtain an approximate and practical solution to real-world systems. Some examples of this arithmetic include the following: implementing algorithms to solve Sudoku puzzles, calculating the factorial of a nonnegative integer, and performing addition and multiplication operations on integers using computer programs.

3) *Interval analysis (IA)*: This category of problems is in essence a combination of infinite arithmetic and finite arithmetic that completely defines a real and semi-ideal problem. Principally, the main purpose of this type of calculation is to ensure that the solution lies within a reasonable confidence interval. IA can also be used as a precondition stage for exploring the initial response to a finite account. Some examples of this arithmetic include the following: estimating the range of possible outcomes for an investment portfolio based

Interval Analysis: Application in the Optimal Control Problems, First Edition. Navid Razmjooy.
© 2024 The Institute of Electrical and Electronics Engineers, Inc.
Published 2024 by John Wiley & Sons, Inc.

on different interest rate scenarios, approximating the solution to a differential equation within a specified error bound, and determining the possible values of a variable given a range of input values and uncertain measurements.

IA is a mathematical framework that focuses on analyzing sets of intervals, treating them as fundamental units instead of singular integers. It is particularly relevant in dealing with uncertainties in physical measurements, integer representations, error analysis, and algorithm stability.

Also known as interval arithmetic, IA allows for the incorporation of system uncertainties by representing values as intervals and examining their influence on system responses [4]. The concept of IA was first introduced by Ramon Moore [5, 6]. While initial interest waned over time, recent years have seen a renewed appreciation for its importance in more realistic models [7].

Researchers from various fields, including mathematics, computer science, and engineering, have recognized that interval arithmetic offers a simple and efficient solution to evaluate the reliability of complex problem responses. Moreover, it accounts for factors that are often overlooked, such as rounding off errors and truncation errors, making IA even more applicable in practical scenarios.

Interval arithmetic finds applications in numerous real-world systems, such as optimal power flow [8], robust control [9–11], antenna design [12], and optimization [13]. Its significance is evident from the fact that the IEEE has established specific standards for IA since 2008, with regular updates – the latest standard being from 2017 [14, 15]. This improved version provides a clearer overview of IA, its historical context, current relevance, and practical applications, while also mentioning the existence of IEEE standards related to this field.

4.2 Introduction to IA

A classical definition of interval numbers within the field of real numbers can be formulated as follows:

$$\mathbb{IR} = \{X \mid X = [\underline{x}, \bar{x}]\}, X = \{x \mid x \in \mathbb{R} \cup \{-\infty, \infty\}, \underline{x} \leq x \leq \bar{x}\}. \tag{4.1}$$

where X represents an interval integer within the set $\mathbb{IR}$ and $\underline{x}$ and $\bar{x}$ denote its lower and upper bounds, respectively.

It is important to note that all interval integers are represented by uppercase symbols, while bold cases indicate vector mode. Additionally, if an interval integer has equal lower and upper bounds, it is considered a degenerate interval integer, i.e., $\mathbb{R} \subset \mathbb{IR}$.

The mid-point value, the width of the interval number, and the radius of an interval integer cap X can be described as follows:

$$x_c = \frac{1}{2}(\overline{x} + \underline{x}), \tag{4.2}$$

$$x_w = \overline{x} - \underline{x}, \tag{4.3}$$

$$x_r = \frac{x_w}{2}, \tag{4.4}$$

Based on the defined properties, an interval integer can be expressed in terms of its center and radius as follows:

$$\begin{aligned} [x] &= x_c + [\Delta x], \\ [\Delta x] &= [-x_r, x_r]. \end{aligned} \tag{4.5}$$

where $[\Delta x]$ is the symmetric interval of $[x]$.

4.3 The Algebra of Interval Sets

The interval subscription of two interval integers $[x], [y] \in \mathbb{IR}$ is defined as follows:

$$[x] \cap [y] = \begin{cases} \varnothing, \text{ if } \overline{x} < \underline{y} \text{ or } \overline{y} < \underline{x} \\ \left[\max\{\underline{x}, \underline{y}\}, \min\{\overline{x}, \overline{y}\} \right] \text{ o.w.} \end{cases} \tag{4.6}$$

It should be noticed that $\varnothing$ is not a closed interval, i.e., $\varnothing \notin \mathbb{IR}$; hence, this relation is partial.

Another partial relation is the interval subset which is defined as follows:

$$[x] \subseteq [y] \text{ if } f \ \underline{y} \leq \underline{x} \text{ and } \overline{y} \geq \overline{x} \tag{4.7}$$

Minkowski is a classical method that extends and generalizes operations such as addition, subtraction, multiplication, and even division to the interval set. This approach allows for a more comprehensive analysis of intervals, as demonstrated below.

4.3.1 Classic Interval Algebra (Minkowski Method)

This algebraic framework extends the fundamental operations of sum $(+)$, subtraction $(-)$, multiplication $(\times)$, and division $(/)$ into the interval set. By assuming $[x] = [\underline{x}, \overline{x}], [y] = \left[\underline{y}, \overline{y}\right]$ and $\circ \in \{+, -, \times, /\}$,

$$[x] \circ [y] = \{x \circ y \in \mathbb{R} \mid x \in [x], y \in [y]\} \tag{4.8}$$

where $0 \notin [y]$. Generally, the main four interval algebra operations are achieved as follows.

$$[x] + [y] = \left[\underline{x} + \underline{y}, \bar{x} + \bar{y}\right] \tag{4.9}$$

$$[x] - [y] = \left[\underline{x} - \bar{y}, \bar{x} - \underline{y}\right] \tag{4.10}$$

$$[x] \times [y] = \left[\min\left\{\underline{xy}, \bar{x}\underline{y}, \underline{x}\bar{y}, \bar{x}\bar{y}\right\}, \max\left\{\underline{xy}, \bar{x}\underline{y}, \underline{x}\bar{y}, \bar{x}\bar{y}\right\}\right] \tag{4.11}$$

By considering $\dfrac{1}{[y]} = \left[\dfrac{1}{\bar{y}}, \dfrac{1}{\underline{y}}\right]$, division can be defined as follows:

$$[x]/[y] = [x] \times \frac{1}{[y]} \tag{4.12}$$

The commutative and associative properties hold for both addition and multiplication as

$$[x] + [y] = [y] + [x], [x] \times [y] = [y] \times [x]$$
$$[x] + ([y] + [z]) = ([x] + [y]) + [z] \tag{4.13}$$
$$[x] \times ([y] \times [z]) = ([x] \times [y]) \times [z]$$

The distributive rules are established based on the following subset:

$$[x] \times ([y] + [z]) \subseteq [x] \times [y] + [x] \times [z]$$
$$([y] + [z]) \times [x] \subseteq [y] \times [x] + [z] \times [x] \tag{4.14}$$

But if we have a real coefficient like x,

$$x \times ([y] + [z]) = x \times [y] + x \times [z]$$
$$([y] + [z]) \times x = [y] \times x + [z] \times x \tag{4.15}$$

It is important to note that in interval analysis, the distributive law differs from that of discrete arithmetic. The order of operations, specifically the arrangement of addition and multiplication, becomes significant in the system.

Let us consider an interval integer $[x]$ and a single functional operation:

1) $[x]([x] + 1) = [-2, 2]$

2) $[x]^2 + [x] = [-1, 2]$

3) $\left([x] + \dfrac{1}{2}\right)^2 - \dfrac{1}{4} = \left[-\dfrac{1}{4}, 2\right]$

As it is clear from the above evaluations, there are three different results for one functional operation, but in infinite arithmetic, these three results are the same. One of the main purposes of the IA is to choose the minimum interval for the system, e.g., $x_w = \overline{x} - \underline{x}$, $x_r = \dfrac{\overline{x} - \underline{x}}{2}$ is the best value for the above evaluations.

4.3.2 Mathematical Norm and Distance in the IA

Size and accuracy criteria in IA are determined based on absolute magnitude. If $[x]$ is represented by $[\underline{x}, \overline{x}]$ and $[y]$ is represented by $[\underline{y}, \overline{y}]$, then we have

$$|[x]| = \max\{|\underline{x}|, |\overline{x}|\} \tag{4.16}$$

$$d([x], [y]) = \max\{|\underline{x} - \underline{y}|, |\overline{x} - \overline{y}|\} \tag{4.17}$$

Please note that due to the inclusion of real numbers ($\mathbb{R}$) within interval numbers ($\mathbb{IR}$), (i.e., $\mathbb{R} \subseteq \mathbb{IR}$), algebraic operations between integers and intervals can be carried out seamlessly.

The distance between an integer x and an interval $[y]$ is also defined as follows:

$$d([x], [y]) = \max\{|x - \underline{y}|, |x - \overline{y}|\} \tag{4.18}$$

By considering the $(w([x]) = \overline{x} - \underline{x})$,

$$w(a[x] + b[y]) = |a| w(|x|) + |b| w([y])$$
$$w([x] \times [y]) = |[x]| w([y]) + |[y]| w([x])$$
$$w\left(\frac{1}{[y]}\right) = \left|\left(\frac{1}{[y]}\right)\right|^2 w([y]) \tag{4.19}$$

Definition 4.1 Interval convergence is defined as the convergence of an interval sequence $[x_k]$ to the interval $[x]$. More formally, we say that the sequence $[x_k]$ converges to $[x]$, if and only if:

$$\forall \varepsilon > 0, \exists N(\varepsilon) > 0, \forall k > N \Rightarrow d([x_k], [x]) < \varepsilon \tag{4.20}$$

which is defined by $\underset{k \to \infty}{\text{Lim}} [x_k] = [x]$ or $[x_k] \to [x]$.

From and numerator section, it can be concluded that the ordinary interval difference (Minkowski difference) cannot result in the true difference or division for a certain value; i.e. $[x] + [-x] \neq 0$ and $\dfrac{[x]}{[x]} \neq 1$.

4.3.3 Kaucher Extended Interval Analysis

Ordinary IA ($\mathbb{IR}$) has certain limitations when it comes to performing arithmetic operations such as addition and division. To address this issue, an extended type of IA called the Kaucher extended IA ($\mathbb{IK}$) was developed. Unlike the ordinary IA, $\mathbb{IK}$

does not require the intervals to have the ascending property, i.e., $\bar{x} > \underline{x}$, which makes it more flexible and efficient for certain types of calculations. The $\mathbb{IK}$ interval arithmetic has been shown to have better characteristics than the IR method [16]. For instance, it provides more accurate and tighter bounds for the solution of nonlinear equations and is less sensitive to rounding errors.

$$[x].[y] = \left[\underline{xy}, \overline{xy}\right], \quad \frac{[x]}{[y]} = [x].\frac{1}{[y]}, \quad \frac{1}{[y]} = \left[\frac{1}{\bar{y}}, \frac{1}{\underline{y}}\right] \tag{4.21}$$

The Kaucher extended interval ($(\mathbb{IK})$) set is divided into three subsets:

1) *Proper interval set*: standard (Minkowski) intervals. This collection is essentially the same as a collection of classic intervals, i.e., $\mathbb{IR}$.
2) *Improper interval set*: Intervals with invertible boundaries, i.e., $\overline{\mathbb{IR}} = \{x = [\underline{x}, \bar{x}] : \underline{x} > \bar{x}\}$.
3) *Degenerate interval set*: This set is an interconnection of the two explained sets, i.e., $\mathbb{IR} \cap \overline{\mathbb{IR}} = \{[\underline{x}, \bar{x}] : \underline{x} = \bar{x}\}$. Here, $\mathbb{IR} = \overline{\mathbb{IR}}$.

4.3.4 Modal Interval Analysis

This method was first introduced by Gardeñes et al. in 1985 [17]. This method can be considered an improved version of the Minkowski classical interval method [18]. The main disadvantage of Minkowski's classical interval calculation is that, for each number of interval operations, when even using accurate calculations, the range of its calculation is often greater than the probable and acceptable approximation.

This issue occurs when some variables are repeated in an equation several times. Consider, for example $[x] = [1, 3]$, the ideal solution for the difference of itself ($[x] - [x]$) and the division into itself ($[x]/[x]$) in this number must be equal in sequence. But, by using the classic IA, we have

$$[x] - [x] = [1, 3] - [1, 3] = [-2, 2] \tag{4.22}$$

$$[x]/[x] = [1, 3]/[1, 3] = \left[\frac{1}{3}, 3\right] \tag{4.23}$$

This phenomenon is known as the *Amplification of dependence* [19].

Modal IA can help address the amplification of the dependence problem that arises in traditional IA. In modal interval calculations, the differential and division operations for two intervals $[x]$ and $[y]$ can be represented as follows:

$$[x] - [y] = [x] - Dual([y]), \tag{4.24}$$

$$[x]/[y] = [x]/Dual([y]), \tag{4.25}$$

where $Dual\left(\left[\underline{y}, \overline{y}\right]\right) = \left[\overline{y}, \underline{y}\right]$.

For instance, for the prior interval,

$$[x] - [x] = [1, 3] - [3, 1] = \{0\}, \tag{4.26}$$

$$[x] - [x] = [1, 3] - [3, 1] = \{0\}, \tag{4.27}$$

Modal IA is a variant of conventional IA that seeks to tackle the issue of increased dependent amplification. The issue of dependency arises when an initial inaccuracy in a particular interval computation is transmitted and magnified in future computations. This phenomenon has the potential to result in excessively cautious approximations and erroneous outcomes. Modal IA addresses the issue of reliance by including the notion of modality, which effectively reflects the intrinsic uncertainty present in the input data. Modal IA involves the use of modal intervals to describe the differential and division operations for two intervals $[x]$ and $[y]$. Modal intervals are characterized as a collection of intervals that exhibit varying modes or degrees of uncertainty. Modal IA offers a more precise depiction of the input data and aids in mitigating the exacerbation of the issue of reliance. The use of this technology enables enhanced and meticulous calculations, proving notably advantageous in scenarios where the supplied data exhibits uncertainty or imprecision.

4.3.5 Hukuhara Difference Method

The previous section highlighted that an ordinary interval difference, also known as the Minkowski difference, cannot provide a correct difference for $[x] + (-[x]) \neq \{0\}$, where $\{0\}$ is a degenerate interval zero. This means that the inverse and opposite of a defined integer are not equal under ordinary differencing.

In 1967, Hukuhara proposed the Hukuhara H-difference as a set $[z]$ such that $[x] \ominus [y] = [z] \Leftrightarrow [x] = [y] + [z]$. The most important feature of this approach is that $[x] \ominus [x] = \{0\}$ [20–23]. The H-difference exists if and only if for $[x] \ominus [y] = [z]$, $[x]$ contains a translate $\{[z]\} + [y]$ of $[y]$.

In 2010, Stefania proposed a generalized version of the H-difference [24]. This approach extended the definition of the H-difference by allowing for different types of intervals, including fuzzy and rough intervals. The generalized H-difference provides a more flexible and powerful framework for interval computations and has found applications in various fields, such as control theory, optimization, and engineering.

Definition 4.2 ([6]) Consider X and Y are two interval values where $[x] = [\underline{x}, \overline{x}]$ and $[y] = [\underline{y}, \overline{y}]$. The gH-difference between these two interval sets can be defined as follows [6]:

$$X \ominus_g Y = Z \Leftrightarrow \begin{cases} \text{(I)} \ X = Y + Z \\ \text{(II)} \ Y = X + (-1)Z. \end{cases} \tag{4.28}$$

4.4 Interval Representations

An interval integer can be described as a backward, forward, or central representation by the following definitions:

$$[x] = \overline{x} + x_w I_B, \tag{4.29}$$

$$[x] = \underline{x} + x_w I_F, \tag{4.30}$$

$$[x] = x_c + x_r I_C. \tag{4.31}$$

where $I_B = [-1, 0]$, $I_F = [0, 1]$, and $I_C = [-1, 1]$ are unit intervals and $\dfrac{1}{Y} = Z$.

Proof: Assuming the fact that $[x] = [\underline{x}, \overline{x}]$, the backward, forward, and central representations for the interval integers are achieved as follows:

$$\left. \begin{array}{l} [x] = \overline{x} + x_w I_b \\ x_w = \overline{x} - \underline{x}, I_b = [-1, 0] \end{array} \right\} \Leftrightarrow [x] = \overline{x} + (\overline{x} - \underline{x})[-1, 0] = [\overline{x}, \overline{x}] + [\underline{x} - \overline{x}, 0] = [\underline{x}, \overline{x}], \tag{4.32}$$

$$\left. \begin{array}{l} [x] = \underline{x} + x_w I_F \\ x_w = \overline{x} - \underline{x}, I_F = [0, 1] \end{array} \right\} \Leftrightarrow [x] = \underline{x} + (\overline{x} - \underline{x})[0, 1] = [\underline{x}, \ \underline{x}] + [0, \overline{x} - \underline{x}] = [\underline{x}, \overline{x}], \tag{4.33}$$

$$\left. \begin{array}{l} [x] = x_c + x_r I_C \\ x_r = \dfrac{\overline{\overline{x}} - \underline{x}}{2}, x_c = \dfrac{\overline{x} + \underline{x}}{2}, \\ I_C[-1, 1] \end{array} \right\} \Leftrightarrow [x] = \dfrac{\overline{x} + \underline{x}}{2} + \dfrac{\overline{x} - \underline{x}}{2}[-1, 1] \tag{4.34}$$

For example,

$$[x] = [1, 4] \rightarrow \begin{cases} X = 4 + 3I_F : \text{Forward Representation} \\ X = 1 + 3I_B : \text{Backward Representation} \\ X = 2.5 + 1.5I_C : \text{Central Representation} \end{cases} \tag{4.35}$$

The basic interval computations between two intervals integers X and Y are given as follows:

$$[x] + [y] = \left\{ \begin{array}{l} [\overline{x} + \overline{y}] + [x_w + y_w]I_F : \text{Forward Representation} \\ [\underline{x} + \underline{y}] + [x_w + y_w]I_B : \text{Backward Representation} \\ [x_c + y_c] + [x_r + y_r]I_C : \text{Central Representation} \end{array} \right\} := [\underline{x} + \underline{y}, \overline{x} + \overline{y}]. \tag{4.36}$$

$$
[x] - [y] = \begin{cases} \frac{1}{2}\left([\bar{x}-\bar{y}] + \left[\underline{x}-\underline{y}\right] + [x_w + y_w](I_B + I_F)\right) = \\ \frac{1}{2}\left([\bar{x}-\bar{y}] + \left[\underline{x}-\underline{y}\right] + [x_w + y_w]I_C\right) : \\ Hybrid\ Representation \\ [x_c - y_c] + [x_r + y_r]I_C : Central\ Representation \end{cases} := \left[\underline{x}-\bar{y},\bar{x}-\underline{y}\right].
$$

$$(4.37)$$

Note that $I_C = I_F + I_B$. As observed, the difference operation in IA utilizes a hybrid forward–backward representation instead of separate forward and backward representations.

This approach allows for performing the same computations with greater efficiency.

$$
[x] \times [y] = \begin{cases} [\overline{xy}] + [\bar{x}y_w + \bar{y}x_w - x_w y_w]I_F : Forward\ Representation \\ \left[\underline{xy}\right] + \left[\underline{x}y_w + \underline{y}x_w + x_w y_w\right]I_B : Backward\ Representation \\ [x_c y_c + x_r y_r] + [x_c y_r + y_c x_r]I_C : Central\ Representation \end{cases} := \left[\min\left\{\underline{xy},\bar{x}\underline{y},\underline{x}\bar{y},\overline{xy}\right\}, \max\left\{\underline{xy},\bar{x}\underline{y},\underline{x}\bar{y},\overline{xy}\right\}\right],
$$

$$(4.38)$$

$$
[x]/[y] = [x] \times \frac{1}{[y]},
$$

$$(4.39)$$

$$
\frac{1}{[y]} = \begin{cases} \left[\frac{1}{\bar{y}}, \frac{y_w}{\bar{y}(y_w - \bar{y})}\right] : Forward\ Representation \\ \left[\frac{1}{\underline{y}}, \frac{-y_w}{\underline{y}\left(y_w + \underline{y}\right)}\right] : Backward\ Representation \\ \left[\frac{y_c}{y_c^2 - y_r^2}, \frac{-y_r}{y_c^2 - y_r^2}\right] : Central\ Representation \end{cases} := \left[\frac{1}{\bar{y}}, \frac{1}{\underline{y}}\right],
$$

$$
0 \notin \left[\underline{y},\bar{y}\right]
$$

For proofing the division operation, we can equate $\frac{1}{[y]} = [z]$, then $[y] \times [z] = 1$

For example, for central representation, we have

$$
\frac{1}{[y]} = [z] \rightarrow \frac{1}{y_c + y_r I_c} = z_c + z_r I_c,
$$

Now by using

$$
[y_c z_c + y_r z_r] + [y_r z_c + y_c z_r]I_c = [1] \rightarrow \begin{cases} y_c z_c + y_r z_r - y_r z_c - y_c z_r = 1, \\ y_c z_c + y_r z_r + y_r z_c + y_c z_r = 1, \end{cases}
$$

Finally, by solving the linear system above,

$$
\begin{cases}
z_c = \dfrac{y_c}{y_c^2 - y_r^2}, \\
z_r = \dfrac{-y_r}{y_c^2 - y_r^2}.
\end{cases}
$$

It is important to note that for multiplication and division involving intervals that include negative and zero integers, they should first be transformed into a positive interval value. This can be achieved by converting the main interval into a negative degenerate value and a positive interval, after which the operations can be performed on them.

In the proposed interval representation, the difference is based on the Minkowski method. However, the Minkowski difference has a significant limitation in that it cannot provide a correct difference for $[x] + (-[x]) \neq \{0\}$, where $\{0\}$ is a degenerate interval zero. This means that, as previously mentioned, the inverse and opposite of a defined integer are not equal under the Minkowski difference. Therefore, there is a need to address this shortcoming and improve the accuracy of interval computations.

To overcome this limitation, alternative interval difference methods have been proposed, such as the Hukuhara difference and its generalized version. These methods have been shown to provide a more accurate representation of the interval difference, particularly for cases involving degenerate intervals and negative values. By using these alternative methods, we can improve the accuracy and reliability of interval computations and make them more suitable for various applications, including engineering, finance, and optimization.

Theorem 4.1 Let $[x] \ominus_F [y]$, $[x] \ominus_B [y]$, and $[x] \ominus_C [y]$ are the forward, backward, and central representations of the gH-difference, such that

i) The forward representation of the gH-difference is $[x] \ominus_g [y] = [x]\ominus_F [y] = \left(\underline{x} - \underline{y}\right) + |x_w - y_w| I_F$,

ii) The backward representation of the gH-difference is $[x] \ominus_g [y] = [x] \ominus_B [y] = (\overline{x} - \overline{y}) + |x_w - y_w| I_B$,

iii) The central representation of the gH-difference is $[x] \ominus_g [y] = (x_c - y_c) + |x_r - y_r| I_C$,

Proof: By replacing $\underline{x}, \underline{y}, \overline{x}, \overline{y}$ with x_w, y_w, x_r, y_r based on equations, the presented representations can be proofed.

Theorem 4.2 The gH-difference for two interval integers $[x] = [\underline{x}, \overline{x}]$ and $[y] = \left[\underline{y}, \overline{y}\right]$ always exists, whereas

$$
[\underline{x}, \overline{x}] \ominus_g \left[\underline{y}, \overline{y}\right] = [\underline{z}, \overline{z}], \tag{4.40}
$$

where $\ominus_g$ can be $\ominus_F$, $\ominus_B$, or $\ominus_C$ and

$$
\underline{z} = \min\left\{\underline{x}-\underline{y}, \bar{x}-\bar{y}\right\}
$$
$$
\bar{z} = \max\left\{\underline{x}-\underline{y}, \bar{x}-\bar{y}\right\} \tag{4.41}
$$

More details on the gH-difference can be found in [6, 24–26]. The following properties were obtained in [24].

Proposition 4.1 Consider κ_c^n be a convex set; then there exist two compact convex sets $([x], [y] \in \kappa_c^n)$ such that [24]

1) $[x] \ominus_g [y] = [x] \ominus [y]$, if gH-difference exists
2) $[x] \ominus_g [x] = [0, 0] = \{0\}$
3) If $[x] \ominus_g [y]$ exists in case (I), $[y] \ominus_g [x]$ exists in case (II) and vice versa,
4) $([x] + [y]) \ominus_g [x] = [y]$
5) $[0, 0] \ominus_g ([x] \ominus_g [y]) = (-[y]) \ominus_g (-[x])$,
6) If $[z] = [0, 0]$ and $([x] \ominus_g [y]) = ([y] \ominus_g [x]) = [z]$ then $[x] = [y]$

The described features can be easily proofed by forward, backward, and central representations.

Please note that when performing multiplication and division operations on intervals that include negative and zero integers, we follow a specific procedure. First, we transform the interval into a positive interval value. This involves representing the main interval as a combination of a negative degenerate value and a positive interval. The multiplication or division operations are then applied to these transformed representations.

4.5 Interval Complex Integers

IA for complex integers, $\mathbb{C}$, similar to the ordinary complex integers is performed in two ways:

a) Cartesian interval arithmetic (rectangles)
b) Polar interval arithmetic (discs)

The operations here are almost the same as the complex number algebras by combining the real interval integers [27].

As central point-based methods hold significant importance in various numerical methods, they serve as the foundation for uncertainty analysis. The central

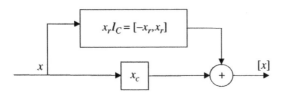

Figure 4.1 Block diagram of the central interval uncertainty.

definition forms the basis for this analysis. The block model of the central model can be represented as follows.

Figure 4.1 displays the block diagram of the central interval uncertainty. where x_c is an integer with the uncertainty that is appeared as $x_r I_C$; therefore, based on the Minkowski method,

$$
\begin{aligned}
& [x] - x_c = x_r I_C, \\
\Rightarrow\; & [x] - x_c = \left[\frac{\overline{x} - \underline{x}}{2}, \frac{x - \overline{x}}{2} \right], \\
\Rightarrow\; & |[x] - x_c| = \left| \left[\frac{\overline{x} - \underline{x}}{2}, \frac{x - \overline{x}}{2} \right] \right| = \left| \frac{\underline{x} - \overline{x}}{2} \right| = |?| \\
\Rightarrow\; & |[x] - x_c| = |[\underline{x}, \overline{x}] - x_c| = |\gamma|
\end{aligned}
\tag{4.42}
$$

Therefore,

$$
|[\underline{x} - x_c, \overline{x} - x_c]| = |\gamma|, \gamma \in [|\underline{x} - x_c|, |\overline{x} - x_c|]
$$

where γ is a constant integer. Hence, we can consider the uncertainty as follows:

$$
|[\underline{x} - x_c]| \leq |\gamma|, |[\overline{x} - x_c]| \geq |\gamma|
\tag{4.43}
$$

Figure 4.2 illustrates the graphical diagram of the uncertainty.

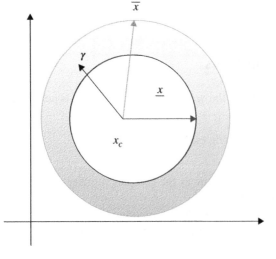

Figure 4.2 Graphical diagram of the uncertainty.

4.6 Interval Vector and Matrix

4.6.1 Vector and Matrix IA

A real-valued interval vector, $[x]$, is a subset of $\mathbb{IR}^n$ which can be taken as the Cartesian multiplicity of n closed intervals. When there is no ambiguity, $[x]$ is simply called an interval vector or interval box that can be described as follows:

$$\left[\vec{x}\right] = [x_1] \times [x_2] \times \ldots \times [x_n],$$
$$[x_i] = [\underline{x}, \overline{x}], i = 1, \ldots, n. \tag{4.44}$$

The ith interval component $[x_i]$ is a projection from $[x]$ in the ith axis.

The empty set from $\mathbb{IR}^n$ must be written as $\emptyset \times \ldots \times \emptyset$, because all of its components are empty. A box sample with two dimensions ($[x] = [x_1] \times [x_2]$) is shown in Figure 4.3.

The lower bound of a box $\boldsymbol{[x]}$ for a vector consists of the lower bounds of its interval components:

$$\underline{x} = \left(\underline{x}_1, \underline{x}_2, \ldots, \underline{x}_n\right)^T \tag{4.45}$$

Similarly, the upper bound for a box $[x]$ of a vector contains the upper bounds of its interval components:

$$\overline{x} = \left(\overline{x}_1, \overline{x}_2, \ldots, \overline{x}_n\right)^T \tag{4.46}$$

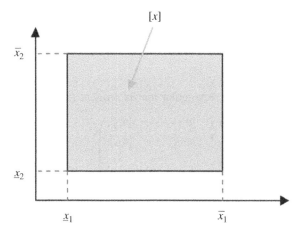

Figure 4.3 A box sample with two dimensions.

Since,

$$\begin{aligned} x_c &\triangleq (x_{c1}, x_{c2}, ..., x_{cn})^T \\ x_r &\triangleq (x_{r1}, x_{r2}, ..., x_{rn})^T \\ x_w &\triangleq \underset{i}{\max} \, (x_{w_i})^T \end{aligned}$$

(4.47)

The addition and subtraction operations are performed element-wise, one component at a time. When it comes to multiplying two vectors, it is considered a dot product utilizing modal analysis techniques.

$$[x] \cdot [y] = ([x_1] \cdot [y_1], [x_2] \cdot [y_2], ..., [x_n] \cdot [y_n])^T$$

(4.48)

4.6.2 Interval Vector Norm

The size and the precision measurements in the vectors are defined by Norm and distance, which are constructed with the absolute magnitude. If $[x] = [\underline{x}, \overline{x}]$ and $[y] = [\underline{y}, \overline{y}]$, then

$$\| [x] \|_1 = \sum_{i=1}^{n} |[x_i]|$$

(4.49)

$$\| [x] \|_\infty = \underset{i}{\max} \, \{ |[x_i]| \}$$

(4.50)

$$d([x], [y]) = \| [x] - [y] \|$$

(4.51)

For each vector x, in the interval $[x]$, $\|x\| \leq \|[x]\|$.

4.6.3 Interval Matrix Analysis

An interval matrix $(m \times n)$ analysis is a rectangular matrix array as follows:

$$[A] = ([a_{ij}])_{1 \leq i \leq m, 1 \leq j \leq n} = \begin{pmatrix} [a_{11}] & [a_{12}] & \cdots & [a_{1n}] \\ [a_{21}] & [a_{22}] & \cdots & [a_{2n}] \\ \vdots & \vdots & \ddots & \vdots \\ [a_{m1}] & [a_{m2}] & \cdots & [a_{mn}] \end{pmatrix}$$

(4.52)

The interval set of matrixes $m \times n$ is illustrated as $\mathbb{ID}^{m \times n}$. The method for illustrating the interval matrices is similar to the interval integers [28].

In the real space, an interval matrix A is illustrated as $[A] = [\underline{A}, \overline{A}]$, where $\underline{A}, \overline{A}$ represents the upper and lower bounds of the interval matrix $[A]$, respectively.

The addition, subtraction, and multiplication operators for interval matrices are defined as noninterval matrices, i.e.,

- if $[A], [B] \in \mathbb{IK}^{m \times n}$, then $[C] = [A] \pm [B]$, where $[a_{ij}] = [b_{ij}] \pm [c_{ij}]$.
- If $A \in \mathbb{ID}^{m \times p}$, then, $A_c = [a_{c(ij)}]$, $A_w = [a_{w(ij)}]$ and $A_r = [a_{r(ij)}]$.
- if $[A], [B] \in \mathbb{ID}^{m \times p}$, then $[C] = [A] \times [B] \in \mathbb{ID}^{m \times n}$ where $[c_{ij}] = \sum_{k=1}^{p} [a_{ik}][b_{kj}]$.

Example 4.1 Consider a real matrix, $A = \begin{pmatrix} 1 & 1 \\ 0 & 1 \end{pmatrix}$ and the vector of the array, $[x] = \begin{pmatrix} [-1, 0] \\ [1, 2] \end{pmatrix}$. In this case, we will have $A.[x] = \begin{pmatrix} [0, 2] \\ [1, 2] \end{pmatrix}$, which indicates that $[0, 2]^T$ belongs to the collection $A. [x]$ and does not belong to the real value, i.e., $\mathbb{B} = \{A.x \mid x \in [x]\}$ as shown in Figure 4.4.

As shown in Figure 4.4, using the ordinary method of multiplication, a relatively high error rate has been obtained, which is known as the *wrapping effect*.

The "wrapping effect" is a phenomenon seen in IA when arithmetic operations are conducted on intervals. The issue at hand is a consequence of the intrinsic lack of accuracy of interval arithmetic, which uses intervals to denote ranges of potential values rather than exact values.

The phenomenon known as the wrapping effect arises when the outcome of a mathematical operation conducted on intervals surpasses the range of values that would be achieved by executing the identical operation on the individual endpoints of those intervals. Stated otherwise, the resultant interval has the potential to exhibit a bigger magnitude than first anticipated, including values that are not included by the original intervals.

4.6.3.1 Conceptional Example: Wrapping Effect
Consider the addition of two intervals $[a, b]$ and $[c, d]$. The wrapping effect can occur if the intervals overlap or if one interval contains the other. In such cases, the resulting interval may extend beyond the range of values obtained by adding

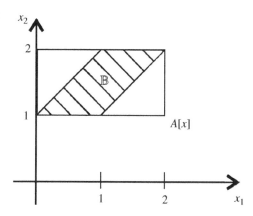

Figure 4.4 An error occurred due to the wrapping effect.

the individual endpoints of the intervals. The wrapping effect can lead to conservative estimates or overestimation of the range of possible values. It is important to be aware of this effect when using IA to ensure accurate and reliable results. Various techniques, such as tightening or refining intervals, can be employed to mitigate the wrapping effect and improve the precision of interval computations.

4.6.4 Interval Matrices Norm

Within the realm of IA, the norm of an interval matrix serves as a quantitative indicator of its scale or extent. The provided data offers insights into the distribution or magnitude of intervals within the matrix. The interval matrix norm is often established by considering the norms of the various intervals included within the matrix. A frequently used method involves the calculation of the maximal norm, which is sometimes referred to as the infinity norm or the supremum norm.

To determine the maximum norm of an interval matrix, it is necessary to compute the maximum norm of each interval element inside the matrix, and afterward identify the highest value among them. This is the maximum distance that may exist between any two points inside the interval matrix.

The determination of the interval matrix norm allows for the acquisition of information on the magnitude and fluctuation of the matrix. The use of interval calculations aids in the comprehension of the effects caused by uncertainties and mistakes, while also offering a means to measure the extent of the intervals implicated. With the help of the absolute magnitudes, the interval matrix norm for an interval integer, $[a]$, can be achieved as follows:

$$\|[A]\|_1 = \sum_j | [a_{ij}] | \tag{4.53}$$

$$\|[A]\|_\infty = \underset{i}{\text{Max}} \sum_{j=1}^{n} | [a_{ij}] | \tag{4.54}$$

For the interval matrix A,

$$[A_r] = \underset{i,j}{\text{max}} \left\{ [a_{r,ij}] \right\} \tag{4.55}$$

$$[A_c] = [a_{c,ij}] \tag{4.56}$$

$$[A_w] = [a_{w,ij}] \tag{4.57}$$

where $a_{r,ij}$, $a_{c,ij}$, and $a_{w,ij}$ are the radius, centered, and width interval values for the a_{ij}, respectively.

4.6.5 Interval Determination of a Matrix

By expanding the Laplace extension, the interval determinant of a matrix can be obtained.

4.6.6 The Inverse of a Regular Interval Matrix

Different methods have been introduced for reversing the interval matrices which have different wrapping effect error values. For a regular real-valued interval matrix A, the inverse matrix $[A]^{-1} = [\underline{B}, \overline{B}]$ can be considered as follows:

$$\underline{B} = \min\{[A]^{-1}\} \tag{4.58}$$

$$\overline{B} = \max\{[A]^{-1}\} \tag{4.59}$$

The order of a regular interval matrix is defined such that all matrices in the interval $[\underline{B}, \overline{B}]$ are nonsingular. This means that the interval determinant does not include degenerate zero values. In other words, for an interval matrix to be regular, none of the determinants of the individual matrices within the interval should be equal to zero. Regular interval matrices are particularly important in IA as they provide reliable and well-behaved solutions. By ensuring that all matrices in the interval are nonsingular, we can avoid the ambiguity and instability that can arise from degenerate cases. The condition of regularity guarantees that the solutions obtained using interval computations are valid and consistent, offering more robust and reliable results when working with interval matrices.

Note that $[A]^{-1}$ is the most compact interval matrix in the $\{[A]^{-1} : A \in [A]\}$. In other words, if the CH is considered convex Hull, $A^{-1} = CH\{[A]^{-1} : A \in [A]\}$.

$[A]^{-1}$ can be explicitly identified for only certain interval matrices and in many cases, it is impossible to calculate the $[A]^{-1}$. So, most times, an estimation of the inverse for the interval matrix has been considered.

4.6.7 Eigenvalues and Eigenvectors of an Interval Matrix

The eigenvalues and eigenvectors of an interval matrix $[A] = [\underline{A}, \overline{A}]$ are equal to

$$\lambda(A) = [\lambda\{A_c\} - \varepsilon, \lambda\{A_c\} + \varepsilon] \tag{4.60}$$

where A_c is the mean value of the matrix, $\lambda\{A_c\}$ is the eigenvalue of the matrix A_c and

$$\varepsilon = \|A[x] - \lambda\{A_c\}[x]\|_2 \tag{4.61}$$

where ε is an interval null vector. Here, x is an eigenvector for the $\lambda\{A_c\}$ and $\|.\|_2$ is the norm-2 of the matrix and can be achieved as follows:

$$
\|[\alpha]\|_2 = \left(\sum_{i=1}^{n} \{\|[\alpha]\|\}^2 \right)^{\frac{1}{2}}
$$
$$
\|[\alpha]\| = \mathrm{col}(\|[\alpha]\|)_{i=1}^{n}
$$
$$
\left|[\alpha]_i\right| = \max\{\|[\underline{\alpha}_i]\|, |[\overline{\alpha}_i]|\}
$$

(4.62)

4.7   Solving Linear Systems with Interval Parameters

Consider a system $Ax = b$. In practice, the values for A and b can be assumed as interval parameters. An interval linear system is shown as follows:

$$
[A][x] = [b]
$$

(4.63)

where $[A]$ is the interval matrix, $[b]$ is the interval vector, and the response of $[x]$ will be also an interval vector.

Generally, the final result can be convex or not. As an example, consider the following interval linear equations:

$$
\begin{pmatrix} [2,4] & [-2,1] \\ [-1,2] & [2,4] \end{pmatrix} x = \begin{pmatrix} [-2,2] \\ [-2,2] \end{pmatrix} \quad \begin{pmatrix} 3.5 & [0,2] & [0,2] \\ [0,2] & 3.5 & [0,2] \\ [0,2] & [0,2] & 3.5 \end{pmatrix} x = \begin{pmatrix} [-1,1] \\ [-1,1] \\ [-1,1] \end{pmatrix}
$$

The final solutions of these two systems with IA are shown in Figure 4.5.

4.8   Interval Functions

4.8.1   Overestimated Interval Function

If a function f is in the real space $\mathbb{R}^n \to \mathbb{R}^m$, its mapping in the interval space $\mathbb{IR}^n \to \mathbb{IR}^m$ will be as follows:

$$
[f([x])] = \{f(x) \mid x \in [x]\}
$$

(4.64)

One of the most important abilities of interval arithmetic is the rapid calculation of the limitation bounds for a function [29].

The interval function $[f([x])]$ is also called the inclusion function of the $f([x])$. This interval may be caused by some overestimations [30]. Consider the general function as follows:

$$
f : \mathbb{R}^n \to \mathbb{R}
$$
$$
(x_1, ..., x_n) \mapsto (f_1(x_1, ..., x_n), f_2(x_1, ..., x_n), ..., f_m(x_1, ..., x_n))
$$

(4.65)

(a)

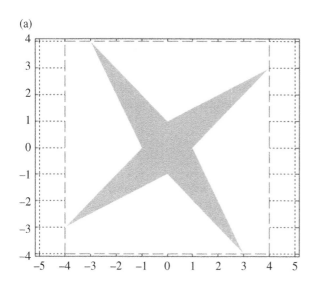

(b)

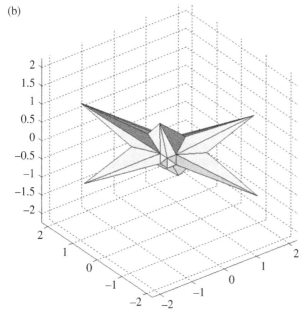

Figure 4.5 The solution region of the (a) two-dimensional and (b) three-dimensional systems in Example 3.7.

where the function f and its components are a finite combination of the main operators, $+,-,\times,\div$, and the main functions (sin, cos, exp, *sqrt*, ...).

By replacing each real variable with an interval variable, any arbitrary interval function $[f_i] : \mathbb{IR}^n \to \mathbb{IR}^m$, can be obtained.

It is important to note that operators or functions should be also replaced with their interval counterparts. This method is known as the *Natural Inclusion Function* [31].

4.8.2 Minimal Interval Function

Interval functions can be defined in different ways. In the general case, the interval function $[f^*]([x])$ has been called minimal, if it has the smallest possible interval for the function f.

$$f([x]) \subseteq [f^*]([x]) \subseteq [f]([x]) \tag{4.66}$$

For further understanding, consider the function $\mathbb{R}^2 \to \mathbb{R}^2$. The interval function $[f]$ of $\mathbb{R}^2 \to \mathbb{R}^2$ is an interval function of f, if

$$\forall [x] \in \mathbb{IR}^n, \ f([x]) \subset [f]([x]) \tag{4.67}$$

Note that the minimal intervals of an interval function are called *Convex Hulls*.

Consider a two-dimensional interval function. Assume function $f \colon \mathbb{R}^2 \to \mathbb{R}^2$, where the variables (x_1, x_2) are mapped into the $[x_1], [x_2]$.

Figure 4.6 illustrates the Venn diagram for mapping a function $f \colon \mathbb{R}^2 \to \mathbb{R}^2$.

When dealing with discontinuous functions, the mapping of $f([x])$ can take on any shape, whether it is convex, nonconvex, or any other form. However, the mapping function f will always guarantee a range of shapes for any existing state. While this range may be larger than the true value of the system, it will always include the exact solution. This is because the interval mapping, $[f]([x])$, approach used in IA provides an enclosure of the solution space, which is guaranteed to contain the true solution.

The shape and size of this enclosure depend on the properties of the function being analyzed and the input intervals. In the case of discontinuous functions,

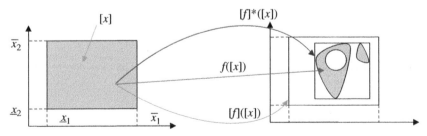

Figure 4.6 Venn diagram for mapping a function $f \colon \mathbb{R}^2 \to \mathbb{R}^2$.

the enclosure may be larger than the true solution space due to the presence of singularities or other nonsmooth features. Despite this limitation, the interval mapping approach provides a powerful tool for analyzing and solving problems involving discontinuous functions. It allows for a rigorous and systematic treatment of uncertainty and variability in the input data and provides a reliable estimate of the solution space. This makes it particularly useful in applications where the input data is uncertain or imprecise, such as in engineering, finance, and science.

4.9 Determining the Minimal Interval

The natural interval functions are not necessarily minimal, as they can be affected by the wrapping effect error. This error causes an inappropriate increase in the unused area of the function, which, in turn, increases the computational cost of the system. The wrapping effect error arises when the boundaries of the input intervals are not aligned with the critical points of the function being analyzed. This can lead to an overestimation of the range of the function and an increase in the computational effort required to obtain accurate results. To address this issue, various methods have been proposed to minimize the wrapping effect error in interval computations. These methods include the use of adaptive subdivision techniques, the introduction of auxiliary functions to capture the behavior of the original function, and the use of nonuniform partitions of the input intervals. By minimizing the wrapping effect error, we can reduce the computational cost of interval computations and obtain more accurate and reliable results. This is particularly important in applications where the cost of computation is a limiting factor, such as in real-time control systems or optimization problems.

4.9.1 Uniform Interval Functions

In these functions, the minimum range of the function is easily determined, e.g., consider the following function:

$$\exp([x]) = [\exp(\underline{x}), \exp(\overline{x})] \tag{4.68}$$

4.9.2 Nonuniform Interval Functions

Given the following unidirectional intervals of the function, these functions require specific algorithms for the evaluation, e.g., to obtain a sinusoidal function of compact nodes,

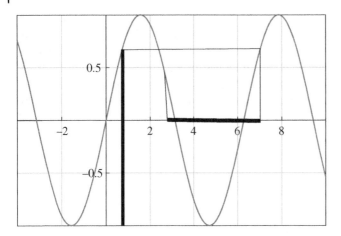

Figure 4.7 Calculation of the value of the box sin([2.6, 7.2]) = [−1, 0.7937].

$$\text{if } \exists kZ \left| 2k\pi - \frac{\pi}{2\epsilon[x]} \right|, \quad then \ \left| \inf([\sin](([x]))) \right| = -1$$

$$\text{else } \inf([\sin]([x])) = \min(\sin \underline{x}, \sin \overline{x})$$

$$\text{if } \exists kZ \left| 2k\pi + \frac{\pi}{2\epsilon[x]} \right|, then \left| \sup([\sin]([x])) \right| = 1 \tag{4.69}$$

$$\text{else } \sup([\sin]([x])) = \max(\sin \underline{x}, \sin \overline{x})$$

where Z represents a set of integers. Figure 4.7 illustrates the calculation for sin ([2.6, 7.2]) = [−1, 0.7937].

4.9.3 Interval Power Series

To extend the IA into the polynomials, interval power is also needed that can be defined as follows:

$$[x]^n \begin{cases} [0, \max(\underline{x}^n, \overline{x}^n)], & n = 2k, 0 \in [x] \\ [\min(\underline{x}^n, \overline{x}^n), \max(\underline{x}^n, \overline{x}^n)], & n = 2k, 0 \notin [x] \\ [\underline{x}^n, \overline{x}^n], & n = 2k + 1 \end{cases} \tag{4.70}$$

where k is a nonnegative integer.

$[f_i]$ is minimal if f_i involves only main operators and functions. Also, if any of the variables $(x_1, ..., x_n)$ are only repeated only once in the equation, $[f_i]$ will be minimal.

Example 4.2 Consider the interval function $f(x) = x^2 - 2x + 1$ in the interval $X = [1, 2]$. Two different solutions have been considered using the natural method for this problem. By calculating the functions of the natural fields, we have

(1) $F_{N1} = (X - 1)^2 = ([1,2] - [1,1])^2 = [0,1]$

(2) $F_{N2} = X^2 - 2X + 1 = [1,2]^2 - 2[1,2] + [1,1] = [-2,3].$

$$(4.71)$$

In this case, the accuracy of the system depends on how the function is selected; as seen, in the first statement, the variable appears only once in the equation, but in the second equation, this variable is repeated twice, which increases inappropriate interval.

Example 4.3 As a practical example, consider the following electrical circuit (Figure 4.8) with practical resistors and variance of $\tilde{R}_1 = \tilde{R}_2 = \tilde{R}_3 = 1^k\Omega \pm 10\%$. Assume that the voltages are degenerate, whereas $V_1 = \{10\}v$, $V_2 = \{5\}v$. In this case,

$$\begin{bmatrix} \tilde{R}_1 + \tilde{R}_2 & -\tilde{R}_2 \\ -\tilde{R}_2 & \tilde{R}_2 + \tilde{R}_3 \end{bmatrix} \begin{bmatrix} I_1 \\ I_2 \end{bmatrix} = \begin{bmatrix} V_1 \\ -V_2 \end{bmatrix}$$

$$(4.72)$$

By using the IA,

$$\begin{bmatrix} \tilde{R}_1 + \tilde{R}_2 & -\tilde{R}_2 \\ -\tilde{R}_2 & \tilde{R}_2 + \tilde{R}_3 \end{bmatrix} \begin{bmatrix} I_1 \\ I_2 \end{bmatrix} = \begin{bmatrix} V_1 \\ -V_2 \end{bmatrix}$$

$$I_1 = \frac{\begin{vmatrix} V_1 & -\tilde{R}_2 \\ -V_2 & \tilde{R}_2 + \tilde{R}_3 \end{vmatrix}}{\begin{vmatrix} \tilde{R}_1 + \tilde{R}_2 & -\tilde{R}_2 \\ -\tilde{R}_2 & \tilde{R}_2 + \tilde{R}_3 \end{vmatrix}}$$

$$(4.73)$$

$$I_2 = \frac{\begin{vmatrix} \tilde{R}_1 + \tilde{R}_2 & V_1 \\ -\tilde{R}_2 & -V_2 \end{vmatrix}}{\begin{vmatrix} \tilde{R}_1 + \tilde{R}_2 & -\tilde{R}_2 \\ -\tilde{R}_2 & \tilde{R}_2 + \tilde{R}_3 \end{vmatrix}}.$$

Figure 4.8 Electrical circuit in the presence of interval uncertainties.

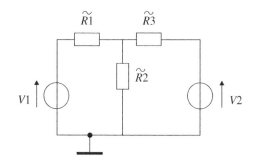

If, for example, we want to calculate the current I_1 in the equation above,

$$(1): I_1 = \frac{V_1(\tilde{R}_2 + \tilde{R}_3) - \tilde{R}_2 V_2}{(\tilde{R}_1 + \tilde{R}_2)(\tilde{R}_2 + \tilde{R}_3) - \tilde{R}_2^2} = [0.45, 0.9],$$

$$(2): I_1 = \frac{V_1\left(1 + \frac{\tilde{R}_3}{\tilde{R}_2}\right) - V_2}{\tilde{R}_1 + \tilde{R}_3 + \frac{\tilde{R}_1 \tilde{R}_3}{\tilde{R}_2}} = [0.42, 0.65]. \tag{4.74}$$

From the above example, it is clear that finding a natural inclusion function can result in different solutions for different definitions. It can be concluded that approach (2) makes a narrower solution than approach (1).

4.10 Interval Derivative and Integral Functions

Interval integrals share significant similarities with ordinary integrals due to their inherent nature of being defined on an interval. Consequently, interval integrals often exhibit lower wrapping effect errors. Conversely, derivatives have instantaneous rates, resulting in higher wrapping effect errors but lower average variations. Dynamic systems encompass both derivatives and integrals.

Thus, in this section, we will delve into the definition and improvements of the derivative and integral operations for interval functions. Several definitions have been proposed for derivatives, and among them, a particularly efficient method that closely aligns with the derivative definition is the approach presented by Stefanio et al., which will be elaborated on in detail in the following explanation [24].

4.10.1 Interval Derivative

Assume the derivative of the function F in the interval X such that

$$F'(X) = \frac{F_w(X)}{X_w} \tag{4.75}$$

Definition 4.3 Assuming $x_0 \in\ ]a, b[$ and h such a way that $x_0 + h \in\ ]a, b[$, then the generalized Hukuhara (gH) derivative of the function $]a, b[\rightarrow \mathbb{IR}$ in x_0 is defined as follows:

$$f'(X_0) = \lim_{h \to 0} \frac{1}{h}\left[f(x_0 + h) \ominus_g f(x_0)\right] \tag{4.76}$$

If $f'(x_0) \in \mathbb{IR}$ satisfies the equation above, f in x_0 will become the generalized Hukuhara derivative.

Definition 4.4 The continuity of an interval function can be expressed as follows:

$$\lim_{x \to x_0} f(x) = f(x_0) \Leftrightarrow \lim_{x \to x_0} \left(f(x) \ominus_g f(x_0) \right) = \{0\} \tag{4.77}$$

Note that the generalized Hukuhara derivative will be satisfied if the function is continuous and the right derivative ($f'_r(x_0)$) and the left derivative ($f'_l(x_0)$) become equal, i.e.,

$$f'_l(x_0) = \lim_{h \nearrow 0} \frac{1}{h} \left[f(x_0 + h) \ominus_g f(x_0) \right] = f'_l(x_0) = \lim_{h \searrow 0} \frac{1}{h} \left[f(x_0 + h) \ominus_g f(x_0) \right] \tag{4.78}$$

Finally, with a central definition, we derive the derivative as follows:

Definition 4.5 Assuming the centered definition of the function ($x = x_c + x_r I_c$) function, we have

$$f'(x_0) = \lim_{h \to 0} \frac{1}{h} \left[f(x_0 + h) \ominus_g f(x_0) \right] = \lim_{h \to 0} \frac{1}{h} \{ [f_c(x_0 + h) - f(x_0)] \}$$
$$+ \left\{ |f_r(x_0 + h) \ominus_g f(x_0)| I_c \right\} \tag{4.79}$$

By the definition above, the partial derivative can be considered as follows:

$$\frac{\partial f(x_0, ..., x_i, ..., x_n)}{\partial x_i} = \lim_{h \to 0} \frac{1}{h} [f(x_0, ..., x_i, ..., x_n + h) \ominus_c f(x_0, ..., x_i, ..., x_n)]$$
$$= \lim_{h \to 0} \frac{1}{h} \{ [f_c(x_0, ..., x_i, ..., x_n + h) - f_c(x_0, ..., x_i, ..., x_n)] \}$$
$$+ \{ |f_r(x_0, ..., x_i, ..., x_n + h) - f_r(x_0, ..., x_i, ..., x_n)| I_c \} \tag{4.80}$$

4.10.2 Interval Integration

Consider the definition of F on the interval X,

$$\int_X F(t) dt \in F(X)(X_w) \tag{4.81}$$

Considering the additive property, with the composite iterative method:

$$\int_X F(t) \mathrm{dt} \in \sum_{i=1}^n F(X)(X_w),$$
$$X = X_1 \cup X_2 \cup ... \cup X_n \tag{4.82}$$

Also, according to [6, 22], for integrating a uniform interval function, assuming that the function $F = \left[\underline{f}, \overline{f}\right]$ is integrable, we have

$$\int_a^b F(t)\Delta t = \left[\int_a^b \underline{f}(t)\Delta t, \int_a^b \overline{f}(t)\Delta t\right] \tag{4.83}$$

If the function is not uniform, the composite integral method is used over uniform intervals.

4.11 Centered Inclusion Method

The best point for evaluating a function on an interval is its center [32]. In the following, two widely used methods of this type have been introduced.

4.11.1 Linearized Interval Functions Around the Center

Assume that $f: \mathbb{R}^n \rightarrow \mathbb{R}$ is a scalar function from a vector $(x_1, ..., x_n)$. Assume that f is differentiable in the interval $[x]$ and x_c is the mean value. The mean value theorem in this case is

$$\forall x \in [x], \exists Z \in [x] \mid f(x) = f(x_c) + g^T(z)(x - x_c) \tag{4.84}$$

where g is the gradient of the function f; with the components $g_i = \dfrac{\partial f}{\partial x_i}$ and $i = 1, ..., n$. As a result,

$$F_T([x]) \doteq f(x_c) + g^T(x_c)([x] - x_c) + \frac{1}{2}([x] - x_c)^T[H]([x] - x_c), \tag{4.85}$$

Assuming that $g = \dfrac{\partial f}{\partial x}$ is an interval function in $[H]([x])$, for each $[x]$ [33]:

$$\forall x \in [x], f(x) \cong f(x_c) + \left[g^T\right]([x])(x - x_c) \tag{4.86}$$

This function can be represented in x as an affine and associated with an indefinite slope dependent on$[f']([x])$. The graph $[f_c]([x])$ can be also illustrated by a hopper with the center of $(x_c, f(x_c))$ to reduce the width of $[x]$. If x_r is getting smaller or if the width becomes zero, then the following happens will be occurred which is rarely in the natural functions:

$$\frac{x_w([f_c]([x]))}{x_w(f([x]))} \rightarrow 1 \tag{4.87}$$

4.11.2 Taylor Inclusion Functions

Considering the centered inclusion method and extending it into the higher derivatives, the Taylor inclusion functions method has been achieved. Consider a twofold extension of the equation [34–36]:

$$F_T([x]) \doteq f(\pmb{x}_c) + g^T(\pmb{x}_c)([x] - \pmb{x}_c) + \frac{1}{2}([x] - \pmb{x}_c)^T[H]([x] - \pmb{x}_c), \qquad (4.88)$$

where $g = \dfrac{\partial f}{\partial x}$ and $[H]([x])$ is the Hessian matrix. The value of $[H]_{ij}$ from $[H]$ is an interval function as follows:

$$h_{ij} = \begin{cases} \dfrac{\partial^2 f}{\partial x_i^2} \text{ if } j = i (i = 1, ..., n), \\ \dfrac{2\partial^2 f}{\partial x_i x_j} \text{ if } j{<}i (i = 1, ..., n), \\ 0 \text{ } OtherWise \end{cases} \qquad (4.89)$$

The symmetric form of the Hessian matrix is $\dfrac{h_{ij} = \partial^2 f}{\partial x_i x_j}$ for all i and j. Therefore, a Taylor single-valued system with order n is given as follows:

$$\begin{aligned} F_T([x]) &\doteq f(\pmb{x}_c) + f'(\pmb{x}_c)([x] - \pmb{x}_c) + ... + f^{(n-1)}(\pmb{x}_c)\frac{([x] - \pmb{x}_c)^{n-1}}{(n-1)!} \\ &+ \left[f^{(n-1)} \right]([x])\frac{([x] - \pmb{x}_c)^n}{(n)!}, \end{aligned} \qquad (4.90)$$

The accuracy of this method is much better than that of previous methods, but the only drawback is that the use of derivatives with different orders makes calculation very complicated and in sensitive cases, because of the presence of noise, the obtained value is wrong [37].

Example 4.4 Consider the function $F(x) = x^3 + e^x$ in the interval $[x] = [0, 1]$. In this case, we have

$$x_c = \frac{1 + 0}{2} = 0.5$$

To evaluate the accuracy of the centered inclusion method, it is important to compare the ranges of the intervals with each other. Ideally, narrower intervals that are closer to the actual range would yield better results. The Hausdorff distance is a commonly used metric for comparing different methods. The Hausdorff distance measures the maximum distance that an object in one set must travel to reach the closest point in another set. It provides a way to quantify the largest possible distance between a point in one set and its nearest neighbor in the opposite set. This distance can be calculated using the following formulation [38–40]:

$$\begin{aligned} H(A, B) &= \max[\mathrm{d}(A, B), \mathrm{d}(B, A)] \\ \mathrm{d}(A, B) &= \max_{a \in A} \mathrm{d}(a, B) \text{ and } \mathrm{d}(a, B) = \min_{b \in B} \mathrm{d}(a, b) = \min_{b \in B} \mathrm{d}|a - b| \end{aligned} \qquad (4.91)$$

Table 4.1 Comparison of the popular inclusion methods.

Method	$F^*(X)$	$F_N(X)$	$F_C(X)$	$F_T(X)$
$X = [0,1]$	[0,3.718]	[1,3.718]	[−1.1,4.63]	[−0.53,4.07]
dist	0 (*Exact value*)	1	1.1	0.35

The results of the solution for the described problem using the centered inclusion method are given in the following:

$$F_N([x]) = [x]^3 + e^{[x]}$$

$$F_C([x]) = F(0.5) + [F']([x])([x] - 0.5)$$

$$F_T([x]) = F(0.5) + F'(0.5)([x] - 0.5) + \frac{1}{2}[f'']([x])([x] - 0.5)^2$$

Table 4.1 illustrates the comparison of the popular inclusion methods.

In the equation above, $F^*(X)$ represents the exact solution. The results demonstrate that the Taylor inclusion method outperforms other methods and provides the most accurate solution. However, Taylor-based methods have a few drawbacks. Firstly, they tend to have longer computation times compared to other methods, which can slow down the process of solving equations. Secondly, due to the inclusion of differentiation in their equations, Taylor-based methods are highly sensitive to high-frequency variations, such as noise. These sensitivity issues can affect the reliability of the solutions obtained using these methods.

4.12 Interval Nonlinear Systems

To determine the equilibrium point of a nonlinear system, the general form of a definite nonlinear system is given as $F(x) = 0$. Various methods have been developed to solve these equations and find the solution for $\dot{x} = F(x)$. Some commonly used methods include the Newton method [41], the Krawczyk method [42], and the backward–forward method [43, 44].

Over time, researchers have dedicated significant efforts to enhance and improve these existing methods. For example, Hensen and Sengupta proposed a method that combines the Gauss–Seidel (GS) and Krawczyk (KC) methods [45]. Neumainer subsequently demonstrated that the Hensen and Sengupta method is more efficient and concise than the Krawczyk method [46]. Benhamou and Jaulin suggested a feed-forward approach based on Waltz's theory [46].

To simplify the solution for the equation $F(x) = 0$, one can consider the Newton–Raphson method [47]. The interval Newton–Raphson method can be implemented by interval linearization using different techniques. One particular approach involves combining interval methods with numerical methods in the following manner:

$$N(X^n) = X_c^n - \nabla F^{-1}\left(X_c^n\right).F\left(X_c^n\right),$$
$$X^{n+1} = N(X^n) \cap X^n \tag{4.92}$$

where $\nabla F^{-1}\left(X_c^n\right)$ is nonsingular interval, $F(.)$ and $\nabla F^{-1}(.)$ are the function and the derivative of the function in a specified point, respectively, and n is the number of the problem-solving iteration.

Example 4.5 Consider the interval function $f(x) = x^3 - 2$ in the interval $x_0 = [1, 3]$.

For solving this nonlinear equation with the Interval Newton–Raphson method, For $n = 0$,

$$N([1,3]) = mid([1,3]) - \frac{(mid([1,3]))^3}{3([1,3])^2} = \left(\frac{1+3}{2}\right) - \frac{\left(\left(\frac{1+3}{2}\right)^3 - 2\right)}{3[1,9]} = \left[0, \left(\frac{16}{9}\right)\right]$$

Therefore, for the first iteration,

$$x_1 = \left[0, \left(\frac{16}{9}\right)\right] \cap [1,3]...$$

After two iterations, the interval will be $[1.25780118, 1.26335408]$. The exact value of the problem is

$$x^2 - 2 = 0 \rightarrow x = \sqrt[3]{2} \approx 1.259921$$

4.13 Analysis of the Interval Dynamic Systems in the Presence of Interval Uncertainties

Differential equations are commonly used to model various dynamical systems in fields such as electrical engineering (e.g., RLC circuits) and mechanical engineering (e.g., pendulums). These systems exhibit dynamics due to the presence of active elements. Therefore, finding suitable methods to solve these systems is crucial in achieving desired objectives, such as system control.

One significant challenge in dynamic systems is dealing with uncertain parameters. These uncertainties can arise from inadequate system modeling, neglecting unknown phenomena or attributes, or unidentified system elements during identification. The analysis of such systems is influenced by the inherent range of uncertainties [45].

Numerous studies have been conducted on dynamic systems. Egiziano et al. [44] proposed special classes of interval methods based on the Taylor series to design robust electromagnetic systems considering uncertainties. The use of the Taylor series in interval systems has also been explored in previous works [48, 49].

In 2011, Wu et al. [50] introduced an interval version of the Euler method for solving ordinary differential equations. However, this method was limited in its ability to handle nonlinear differential equations.

Addressing dynamical problems under interval uncertainties is more challenging compared to static problems. When applying interval methods to solve interval differential equations, the high iteration involved increases the likelihood of the wrapping effect, which may even lead to divergence in numerical methods. Therefore, besides the interval method itself, it is important to consider a compressed interval by modifying overestimations.

In this section, an improved method for solving differential equations with interval uncertainty is proposed. This methodology combines numerical methods with IA, providing a comprehensive approach to handle the challenges posed by interval uncertainties.

4.13.1 Solving the Interval Initial Value Problems

Consider the following interval initial value problems:

$$
\begin{cases}
\dot{X} = F(t, X, \Delta), a \le t \le b, \\
X(t_0) = X_0.
\end{cases}
\tag{4.93}
$$

where $F : [a, b] \times \mathbb{IR}^m \times \mathbb{IR}^k \to \mathbb{IR}^m$ and

$$
F(t, X, \Delta) = \left[\underline{f}(t, X), \overline{f}(t, X) \right], \text{ for } X \in \mathbb{IR}^m
$$
$$
X = [\underline{x}, \overline{x}], X_0 = [\underline{x}_0, \overline{x}_0] \text{ and } \Delta = [\underline{\Delta}, \overline{\Delta}]
\tag{4.94}
$$

where **F**, **X**, and **Δ** are interval vectors. It is also assumed that the initial conditions and the coefficients of the system are real-valued and interval.

Due to the continuous nature of the differential operator, direct utilization in computer programming is not feasible. Therefore, numerical methods are employed to approximate the behavior of the differential operator through discrete differential operators.

One commonly used iterative numerical method for solving differential equations is the Runge–Kutta method (RKM). In this section, we aim to extend this method to an interval space, enabling its application in solving differential equations with interval uncertainties. This extension facilitates the incorporation of IA into the numerical solution process, thereby enhancing the handling of uncertainties inherent in the system.

Theorem 4.3 If the function F in a box $[B]$ contains a continuous variable X_0, has an interval uncertainty Δ in the variable X, and satisfies the Lipschitz conditions, i.e.,

$$\exists L \forall X_1, X_2 \in [B], \|(F(X_1, t, ?) - F(X_1, t, \Delta))\| \leq L\|X_1\text{-}X_2\| \tag{4.95}$$

The solution of Eq. (4.95) still be included in the box $[B]$.

Argument 4.1 By integrating Eq. (4.95),

$$X(t) = X_0(t) + \underbrace{\int_{t_0}^{t_f} F(X(\tau), \tau, \Delta)d\tau}_{T_\Delta(X(t))} \tag{4.96}$$

Assuming the coefficients of the vector function F are unique, we can incorporate interval uncertainties into the dynamic system.

$$X(t) = X_0(t) + \underbrace{\left[\int_{t_0}^{t_f} \underline{f}(X(\tau), \tau)d\tau, \int_{t_0}^{t_f} \overline{f}(X(\tau), \tau)d\tau\right]}_{T_\Delta(X(t))} \tag{4.97}$$

where $T_\Delta(X(t))$ is a mapping in the Banach space, which, according to the Lipschitz conditions, the condition of the fixed point is satisfied.

Theorem: If F is an interval compactness and $L \leq K \leq 1$, the problem solution will be unique. Also, the iterative sequence converges the following solution:

$$X_{n+1}(t) = X_0(t) + \left[\int_{t_0}^{t_f} \underline{f}([X(\tau)]^n, \tau)d\tau, \int_{t_0}^{t_f} \overline{f}([X(\tau)]^n, \tau)d\tau\right] \tag{4.98}$$

Argument 4.2 The uniqueness of the solution can be proved by the proof by contradiction.

For convergence,

$$\begin{aligned}
\|X_{n+1}(t) - X(t)\| &\leq k\|X_n(t) - X(t)\| \\
\|X_{n+1}(t) - X(t)\| &\leq k^2\|X_{n-1}(t) - X(t)\| \\
&\vdots \\
\|X_{n+1}(t) - X(t)\| &\leq k^n\|X_0(t) - X(t)\|
\end{aligned} \tag{4.99}$$

Based on the assumption, for large quantities n, $k^n \ll 1$; therefore, we can conclude that

$$X_{n+1}(t) \rightarrow X(t) \tag{4.100}$$

4.14 The Interval Runge–Kutta Method (IRKM) for Interval Differential Equations

4.14.1 Introduction

In the following, boldface letters will describe vector intervals. Consider an interval-valued initial value problem (IVP) as follows:

$$\begin{cases} \dot{X} = F(t, X, \Delta), a \leq t \leq b, \\ X(t_0) = X_0. \end{cases} \tag{4.101}$$

with $F: [a, b] \times I(\mathbb{R}^m) \times I(\mathbb{R}^k) ? I(\mathbb{R}^m)$ and

$$F(t, X, \Delta) = \left[\underline{f}(t, X, \Delta), \overline{f}(t, X, \Delta) \right], \quad \text{for } X \in I(\mathbb{R}^m),$$
$$X = [\underline{x}, \overline{x}], \quad X_0 = \left[\underline{x_0}, \overline{x_0} \right] \text{ and } \Delta = \left[\underline{\Delta}, \overline{\Delta} \right] \tag{4.102}$$

where F, X, and Δ are interval vectors.

Suppose the initial conditions and coefficients of the system are defined as real intervals. The system can be either linear or nonlinear. Since differentiation is a continuous operation, directly implementing it in computer programming is not feasible. To address this, numerical methods are employed to discretize the differentiation process.

One popular class of iterative numerical techniques used for solving ordinary differential equations is the RKM. In this study, we aim to extend the application of RKMs to solve equations with interval uncertainties. This extension allows us to incorporate IA into the numerical solution process, enhancing our ability to handle uncertainties present in the system.

Theorem 4.4   Consider Eq. (4.101). Assume $\underline{x}_i$ and $\overline{x}_i$ are real-valued functions such that $\underline{x}_i \leq \overline{x}_i$, $\underline{x} = \begin{pmatrix} \underline{x}_1 \\ \vdots \\ \underline{x}_m \end{pmatrix}$, $\overline{x} = \begin{pmatrix} \overline{x}_1 \\ \vdots \\ \overline{x}_m \end{pmatrix}$ and for all $t \in [a, b]$ and $F: [a, b] \times$ $I(\mathbb{R}^m) \times I(\mathbb{R}^k) \rightarrow I(\mathbb{R}^m)$ be continuous. Let X be the interval-valued function defined by $X = \left[\underline{X}, \overline{X} \right]$. If F satisfies the Lipschitz condition, i.e.,

$$D(F(t, X), F(t, Z)) \leq LD(X, Z),$$
$$\forall (t, X) \in [a, b] \times I\mathbb{R}^m. \tag{4.103}$$

By considering $x_{-i}^n(t) = x_{-i}(t_{n-1} + h)$ and $x_{-i}^{n+1}(t) = x_{-i}(t_{n-1} + h)$, where h is the horizon, the unique solution to the inclusion problem can be achieved by an extension of the RKMs and Hukuhara approach, *IRKM* is achieved:

$$(I)\underline{x}_i^n(t) \leq \overline{x}_i^n(t), \begin{cases} \underline{x}_i^{n+1}(t) = \overline{x}_i^n(t) + h\underline{\Phi}_i, \\ \overline{x}_i^{n+1}(t) = \overline{x}_i^n(t) + h\overline{\Phi}_i. \end{cases} \tag{4.104}$$

$$(II)\underline{x}_i^n(t) \geq \overline{x}_i^n(t), \begin{cases} \underline{x}_i^{n+1}(t) = \overline{x}_i^n(t) + h\overline{\Phi}_i, \\ \overline{x}_i^{n+1}(t) = \overline{x}_i^n(t) + h\underline{\Phi}_i. \end{cases} \tag{4.105}$$

where

a) If we solve the problem with the first-order IRKM$^{\text{1st}}$,

$$\begin{aligned} \underline{\Phi} &= \underline{f}(t, X(t), \Delta), \\ \overline{\Phi} &= \overline{f}(t, X(t), \Delta). \end{aligned} \tag{4.106}$$

b) If the purpose is to solve the inclusion problem with the second-order IRKM$^{\text{2nd}}$,

$$\begin{aligned} \underline{\Phi} &= \underline{f}\left(t + \frac{h}{2}, X\left(t + \frac{h}{2}f(t, X(t))\right), \Delta\right), \\ \overline{\Phi} &= \overline{f}\left(t + \frac{h}{2}, X\left(t + \frac{h}{2}f(t, X(t))\right), \Delta\right). \end{aligned} \tag{4.107}$$

c) If the purpose is to solve the inclusion problem with the fourth-order IRKM$^{\text{4th}}$,

$$\begin{aligned} \underline{\Phi} &= \frac{1}{6}(\underline{K}_1 + 2\underline{K}_2 + 2\underline{K}_3 + \underline{K}_4), \\ \overline{\Phi} &= \frac{1}{6}(\overline{K}_1 + 2\overline{K}_2 + 2\overline{K}_3 + \overline{K}_4). \end{aligned} \tag{4.108}$$

where

$$\begin{cases} \underline{K}_1 = \underline{f}(t_n, X^n(t), \Delta), \\ \underline{K}_2 = \underline{f}\left(t_n + \frac{h}{2}, X^n(t) + \frac{\underline{K}_1}{2}, \Delta\right), \\ \underline{K}_3 = \underline{f}\left(t_n + \frac{h}{2}, X^n(t) + \frac{\underline{K}_2}{2}, \Delta\right), \\ \underline{K}_4 = \underline{f}(t_n + h, X^n(t) + \underline{K}_3, \Delta), \end{cases} \quad \begin{cases} \overline{K}_1 = \overline{f}(t_n, X^n(t), \Delta), \\ \overline{K}_2 = \overline{f}\left(t_n + \frac{h}{2}, X^n(t) + \frac{\overline{K}_1}{2}, \Delta\right), \\ \overline{K}_3 = \overline{f}\left(t_n + \frac{h}{2}, X^n(t) + \frac{\overline{K}_2}{2}, \Delta\right), \\ \overline{K}_4 = \overline{f}(t_n + h, X^n(t) + \overline{K}_3, \Delta). \end{cases} \tag{4.109}$$

Note that this procedure can be applied to all of IRKMs.

Proof:

a) For more understanding, let us utilize the IRKM$^{\text{1st}}$ (which can be considered an inclusion extension of the well-known routine Euler method).

Assume $F(t, X, \Delta)$ is a monotonic function and $X \in \mathbb{IR}^m$. The first step is to solve a second-order differential equation by using the Euler method. Suppose that $X_{n+1}(t)$ is a unique solution of the equation with the interval real positive integer N with the points $\{t_0, t_1, ..., t_N\}$, where $t_{n+1} = t_n + h$ and $h = \frac{b-a}{N}$ is the horizon or the same step size. Now, consider the problem in Eq. (4.101). By extending the equation in [51], the interval Euler equation for this problem can be formulated as follows:

$$X_i^{n+1}(t) = X_i^n(t) + hF\left(t, X_i^n(t), \Delta\right) \qquad (4.110)$$

To get better results, we utilized the Hukuhara difference for extending the formula. Here, the forward representation for Hukuhara differencing is utilized (see Eq. (4.98)), and two solutions have been achieved for Eq. (4.19). For the first term (I) $\underline{x}_{n+1}(t) \le \overline{x}_{n+1}(t)$:

$$X_i^{n+1}(t) \Theta_F X_i^n(t) = hF\left(t, X_i^n(t), \Delta\right) \qquad (4.111)$$

And for the left-hand side, we have

$$
\begin{aligned}
X^{n+1}(t) \Theta_F X^n(t) &= \left[\underline{x}^{n+1}(t) + x_w^{n+1} I_F\right] \Theta_F \left[\underline{x}^n(t) + x_w^n I_F\right] \\
&= \left[\underline{x}^{n+1}(t) + x_w^{n+1} I_F\right] - \left[\underline{x}^n(t) + x_w^n I_F\right]
\end{aligned}
\qquad (4.112)
$$

where Θ_b is the forward representation of the generalized differencing.

By considering $I_F = [0, 1]$ and interval multiplying with $y_{(n+1)w}$ and $y_{(n)w}$:

$$
\begin{aligned}
X^{n+1}(t) \Theta_F X^n(t) &= \left[\underline{x}^n(t) + \overline{x}^n(t)\right] + \left[0, \overline{x}^{n+1}(t) - \underline{x}^{n+1}(t) + \underline{x}^n(t) - \overline{x}^{n+1}(t)\right] \\
&= \left[\underline{x}^{n+1}(t) - \underline{x}^n(t) + \overline{x}^{n+1}(t) - \overline{x}^n(t)\right].
\end{aligned}
\qquad (4.113)
$$

By equivalence this result by the right-hand side of Eq. (4.29),

$$(I) \begin{cases} \underline{x}^{n+1}(t) = \underline{x}^n(t) + h\underline{f}(t, X^n(t), \Delta), \\ \overline{x}^{n+1}(t) = \overline{x}^n(t) + h\overline{f}(t, X^n(t), \Delta). \end{cases} \qquad (4.114)$$

For the second form, we have

$$X^n(t) \Theta_F X^{n+1}(t) = (-1)hF(t, X^n(t), \Delta) \qquad (4.115)$$

And for the left-hand side, we have

$$X^n(t) \Theta_F X^{n+1}(t) = \left[\underline{x}^n(t) + \underline{x}^n(t) I_F\right] \Theta_F \left[\underline{x}^{n+1}(t) + \underline{x}_w^{n+1}(t) I_F\right]. \qquad (4.116)$$

After solving the left-hand side equation and putting equal it with the right-hand side like term (I), $\underline{x}^n(t, h) \geq \overline{x}^n(t, h)$ can be described as follows:

$$(II)\begin{cases} \underline{x}^{n+1}(t) = \underline{x}^n(t) + h\overline{f}(t, X^n(t), \Delta), \\ \overline{x}^{n+1}(t) = \overline{x}^n(t) + h\underline{f}(t, X^n(t), \Delta). \end{cases} \qquad (4.117)$$

This proof can be applied to the parts (*b*), (*c*), and IRKM^{nth} with any order of *n*. Now we develop the interval monotonic problem solution. Let us study the problem as an extension of real-valued function $F(t, X, \Delta) = \left[\underline{f}.\overline{f}\right]$. We supposed that F is monotonic. So there are two different cases [52]:

Case 1) If the components of F are increasing functions:

$$\underline{x}_i^n(t) \leq \overline{x}_i^n(t), \qquad\qquad \underline{x}_i^n(t) \geq \overline{x}_i^n(t),$$

$$(I)\begin{cases} \underline{x}_i^{n+1}(t) = \underline{x}_i^n(t) + h\underline{\Phi}_i, \\ \overline{x}_i^{n+1}(t) = \overline{x}_i^n(t) + h\overline{\Phi}_i. \end{cases} \qquad (II)\begin{cases} \underline{x}_i^{n+1}(t) = \underline{x}_i^n(t) + h\overline{\Phi}_i, \\ \overline{x}_i^{n+1}(t) = \overline{x}_i^n(t) + h\underline{\Phi}_i. \end{cases}$$

$$(4.118)$$

Case 2) If the components of F are decreasing functions:

$$\underline{x}_i^n(t) \leq \overline{x}_i^n(t), \qquad\qquad \underline{x}_i^n(t) \geq \overline{x}_i^n(t),$$

$$(I)\begin{cases} \underline{x}_i^{n+1}(t) = \underline{x}_i^n(t) + h\overline{\Phi}_i, \\ \overline{x}_i^{n+1}(t) = \overline{x}_i^n(t) + h\underline{\Phi}_i. \end{cases} \qquad (II)\begin{cases} \underline{x}_i^{n+1}(t) = \underline{x}_i^n(t) + h\underline{\Phi}_i, \\ \overline{x}_i^{n+1}(t) = \overline{x}_i^n(t) + h\overline{\Phi}_i. \end{cases} \qquad (4.119)$$

4.14.2 Generalized IRKM (GIRKM) Based on Switching Points

As noted above, the equations will be true as long as the problem is monotonic. If the problem is not monotonic, there will be no definite solution to classify it into decreasing or increasing problems. This type of problem has some critical points, where the solution among them satisfies one of the decreasing or increasing conditions. Here, based on [24], they have been called switching points. As a result, the method will have a switch among the critical points. In simple terms, at first, the function will be divided into some monotonic subfunctions, and then, each subfunction will be solved based on its monotonic position based on Eqs. (4.118) and (4.119).

For instance, consider the function $f(x) = x^2 - \cos(x)$. This function has a critical point at about $x_{critical} = 0$. From the figure, it is clear that the function is divided into two parts where the points less than the critical point comprise a decreasing

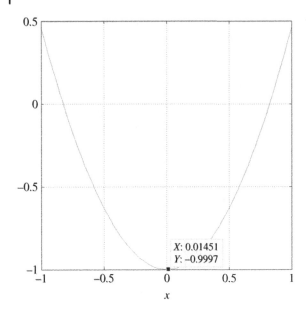

Figure 4.9 Plot of the function $f(x) = x^2 - \cos(x)$ and its critical point.

function, and those with greater values from the critical point comprise an increasing function (see Figure 4.9).

From here, the horizon, h is dynamic and nonconstant. By considering one switching point in the problem above and solving it with the proposed GIRKM,

$$X_i^{n+1}(t)\Theta_F X_i^n(t) = \left[X_i^{n+1}\Theta_F X_i^{cp}\right] + \left[X_i^{cp}\Theta_F X_i^n\right], \tag{4.120}$$

$$X_i^n(t)\Theta_F X_i^{n+1}(t) = \left[X_i^n T_F X_i^{cp}\right] + \left[X_i^{cp} T_F X_i^{n+1}\right]. \tag{4.121}$$

where $X_i^{cp} = \left[X_i^{cp(1)}, ..., X_i^{cp(k)}\right]^T$ is the function in the critical (switching) point. A generalized approach for problems with n switching points can be solved as follows:

$$X_i^{n+1}(t)\Theta_F X_i^n(t) = \left[X_i^{n+1}\Theta_F X_i^{cp(1)}\right] + \sum_{i=1}^{j}\left[X_i^{cp1} T_F X_i^{cp(i+1)}\right] + \left[X_i^{cp(j+1)} T_F X_i^k(t)\right]. \tag{4.122}$$

$$X_i^n(t)\Theta_F X_i^{n+1}(t) = \left[X_i^n(t)\Theta_F X_i^{cp(1)}\right] + \sum_{i=1}^{j}\left[X_i^{cp1}\Theta_F X_i^{cp(i+1)}\right] + \left[X_i^{cp(j+1)}\Theta_F X_i^k(t)\right]. \tag{4.123}$$

Finally, for refining the proposed method, a subdivision approach is employed [27]. Subdivide an interval vector $X = (X_1, ..., X_n)$ where N_i illustrates the number of critical points:

$$X_i = \cup_{j=1}^{N_i} x_{ij},$$

$$N_i = \sum_{i=1}^{n} CP(X_i). \qquad (4.124)$$

where $CP(f(i))$ describes the critical points.

The pseudocode of the proposed method is summarized as follows:

Input:

$h, T = [a, b],$

$\Delta_K = [\underline{\Delta}, \overline{\Delta}]$, (K: number of uncertainties)

$X_j = [\underline{x}_j, \overline{x}_j], j = 1, 2, ..., m$ (m: number of states)

$\mathbf{F}(t, \mathbf{X}(t))$

$$\Phi = f\left(t, X_i^n(t), \Delta\right),$$

Output:

$$X$$

Start

Divide the interval into two-fold intervals: $T = [a, c] \cup [c, b],$
Apply the following operations in two separate intervals:

if X is monotonic
if X is increasing

$$if : \underline{x}_i^n(t) \leq \overline{x}_i^n(t), \begin{cases} \underline{x}_i^{n+1}(t) = \underline{x}_i^n(t) + h\underline{\Phi}_i, \\ \overline{x}_i^{n+1}(t) = \overline{x}_i^n(t) + h\overline{\Phi}_i. \end{cases}$$

$$if : \underline{x}_i^n(t) \geq \overline{x}_i^n(t), \begin{cases} \underline{x}_i^{n+1}(t) = \underline{x}_i^n(t) + h\overline{\Phi}_i, \\ \overline{x}_i^{n+1}(t) = \overline{x}_i^n(t) + h\underline{\Phi}_i. \end{cases}$$

else-if Y is decreasing

$$if : \underline{x}_i^n(t) \leq \overline{x}_i^n(t), \begin{cases} \underline{x}_i^{n+1}(t) = \underline{x}_i^n(t) + h\overline{\Phi}_i, \\ \overline{x}_i^{n+1}(t) = \overline{x}_i^n(t) + h\underline{\Phi}_i. \end{cases}$$

$$if : \underline{x}_i^n(t) \geq \overline{x}_i^n(t), \begin{cases} \underline{x}_i^{n+1}(t) = \underline{x}_i^n(t) + h\underline{\Phi}_i, \\ \overline{x}_i^{n+1}(t) = \overline{x}_i^n(t) + h\overline{\Phi}_i. \end{cases}$$

end-if

end-if
else-if X is not monotonic
find switching points as

$$N_i = \sum_{i=1}^{n} CP(X_i),$$

for all switching points apply monotonic operations and then,

$$X_i = \cup_{j=1}^{N_i} X_{ij},$$

end-if
union separate intervals.
end

In [50], a simple Euler method was proposed to solve the interval linear IVP problems which have a restriction, i.e., when there is nonlinearity in the differential equation nature, the method of [50] fails. Here, the method is improved based on the Hukuhara method and RK-based methods to provide a successful solution for the linear and nonlinear systems based on interval arithmetic.

4.14.3 Numerical Examples

Case Study 4.1   Vehicle dynamic analysis

For instance, assume we want to solve the equation below [53]:

$$\ddot{X} = F(t, X, \dot{X}, \Delta) = A_0 + A_1 X + A_2 \dot{X} \tag{4.125}$$

where the coefficients A_0, A_1 and A_2 are initial conditions. $X_1 = [X_0], X_2 = [\dot{X}_0]$ are real intervals. Suppose that the parameters A_0, A_1 and A_2 are time independent.

It is important to note that when dealing with differential equations of order greater than 1, it is necessary to transform them into a first-order dynamic system to simplify the implementation of IRKMs. This transformation allows for a more straightforward application of IRKMs in solving such equations.

$$\begin{cases} \dot{X}_1 = X_2, \\ \dot{X}_2 = A_0 + A_1 X_1 + A_2 X_2, \end{cases} \tag{4.126}$$

And initial conditions, $X_1(0) = [X_0]$, $X_2(0) = [\dot{X}_0]$ are real intervals. Since the IRKM can be used to solve the problem as follows:

$$\begin{cases} X_1^{n+1}(t) = X_2^n(t) + hF(t, X_2^n(t), \Delta) \\ X_2^{n+1}(t) = X_2^n(t) + hF(t, X_1^n(t), X_2^n(t), \Delta) \end{cases} \tag{4.127}$$

We assume that the initial values $X_1(0) = X_2(0) = 0$ are degenerate zero intervals.

For analyzing the accuracy of the methods, we can compare their central value of them with the exact method. Since, assuming $A_0 = 100$, $A_1 = -10$ and $A_2 = -1$ in the time interval $t \in [0, 3]$ within 0.1 step size (h), the final solution can be considered as follows:

Figure 4.10 shows the comparison of the central value for the RK methods with the exact solution.

Based on the results, it is evident that the RKM^{1st} exhibits a higher approximation error compared to RKM^{2nd} and RKM^{4th}. This discrepancy is expected and highlights that using the first-order RKM may lead to incorrect intervals for system modeling. It is important to note that increasing the order of RKMs also leads to an increase in solution complexity and elapsed time. Therefore, when choosing a method, it is crucial to consider the desired balance between accuracy and speed. In conclusion, it is not recommended to solely rely on the first-order RKM.

Now, by using interval arithmetic and assuming $A_0 \in [99,101]$, $A_1 \in [-10.1, -9.9]$, $A_2 \in [-1.01, -0.99]$ with the previous initial value, time interval, and step size, the solution is given in Figure 4.11.

For a deeper understanding, the interval values are also gathered in Table 4.2.

It is clear from Figure 4.12 that except the IRKM^{1st} which has some errors in determining the interval, the IRKM^{2nd} and IRKM^{4th} give a reasonable interval for the system model solution.

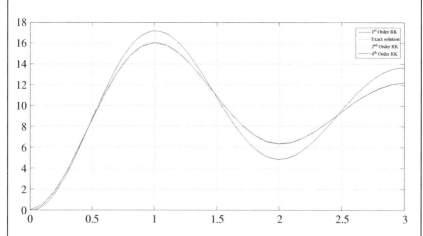

Figure 4.10 Comparison of the central value for the RK methods with the exact solution.

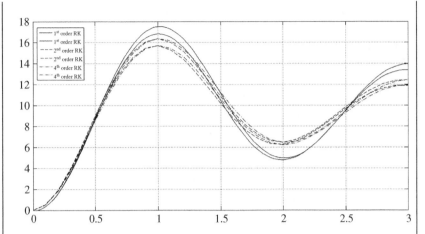

Figure 4.11 The solution of the proposed IRKMs for Case Study 4.1.

Table 4.2 Comparison of the interval RK methods.

| Time | IRKM^{1st} [53] | | IRKM^{2nd} | | IRKM^{4th} | |
	Lower bound	Upper bound	Lower bound	Upper bound	Lower bound	Upper bound
0.5	8.63	8.85	8.79	9.01	8.56	8.79
1	16.81	17.52	15.68	16.38	15.71	16.38
2	4.79	4.95	6.28	6.48	6.23	6.46
2.8	12.77	13.12	11.66	12.05	11.49	11.84
3	13.37	13.96	11.9	12.46	11.95	12.46

As mentioned in the previous section, the proposed IRKM has mainly two parts: The first part is to obtain a step size and a prior enclosure of the solution, which includes the exact solution of the period from the current time step to the next time step and the second part is to achieve a narrower enclosure of the solution [54].

Since a bisection technique is applied to this problem to analyze the results. To do so, the interval vector $X = (X_1, ..., X_n)$ is divided into two subsections (Bisection technique). Since,

$$X = U_{j=1}^{2}(X_j) \tag{4.128}$$

In other words, the system is divided into two parts; each part is solved by interval arithmetic and the general solution is obtained from the union of these two solutions at each point. The results are shown in Figure 4.13.

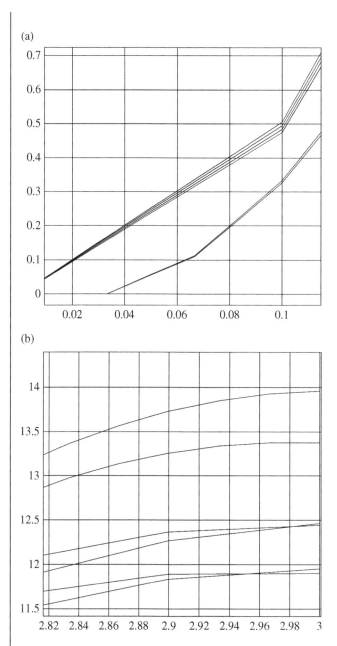

Figure 4.12 The focused result of the proposed IRKMs at the beginning and finishing. The lines are IRKM[1st], the dashed lines are IRKM[2nd] and the point-dash lines describe the IRKM[4th] at (a) the beginning and (b) the finishing.

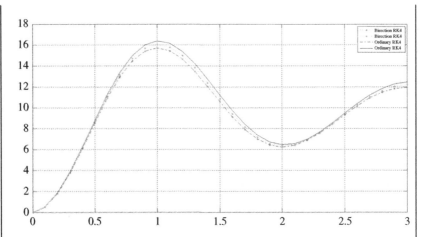

Figure 4.13 Ordinary and developed IRKM4th on the problem.

Table 4.3 Comparison of the ordinary and bisection fourth-order RK methods.

	Ordinary IRKM^{4th}			Bisection IRKM^{4th}		
Time	Lower bound	Upper bound	Interval space	Lower bound	Upper bound	Interval space
0.5	8.56	8.796	0.236	8.56	8.678	0.118
1	15.71	16.38	0.67	15.71	16.05	0.34
2	6.236	6.461	0.225	6.236	6.346	0.11
2.8	11.49	11.84	0.35	11.49	11.67	0.18
3	11.95	12.46	0.51	11.95	12.21	0.26
$\sqrt{(\sum(Intervalspace))^2} =$			0.9684	$\sqrt{(\sum(Intervalspace))^2} =$		0.4916

It is clear from Figure 4.5 that the obtained interval of the bisection method provides a narrower interval than the ordinary method.

Table 4.3 indicates the comparison of the ordinary and bisection 4th order RK methods.

Table 4.4 provides a detailed comparison, revealing that the bisection method exhibits a narrower interval space (upper bound − lower bound) compared to the ordinary method. Additionally, the total norm of the bisection method is lower than that of the ordinary method. These findings demonstrate that the bisection method produces a narrower interval. By computing the critical points for X_1 and X_2 (see Figure 4.14), the position of switching points is obtained as follows:

Table 4.4 Describing the switching points of the solution for Case Study 4.1.

	Increasing	Decreasing
X_1	[0, 1] and [2, 3]	[1, 2]
X_2	[0, 0.5] and [2.5, 3]	[1.5, 2.5]

Therefore, the generalized IRKM^{4th} (*GIRKM*^{4th}) can be shown as follows:

(a)   (b)

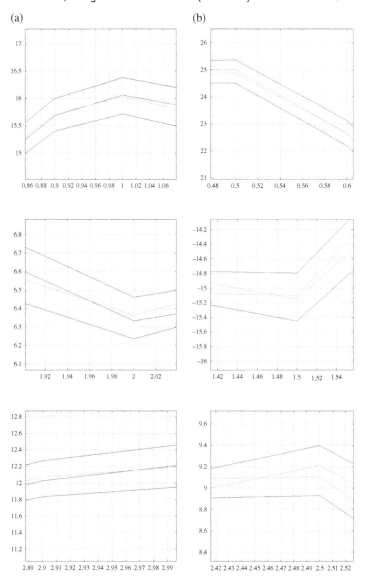

Figure 4.14 Switching points of the solution for (a) X_1, (b) X_2.

Figure 4.15 shows the ordinary and developed IRKM^{4th} on problem by considering the switching points.

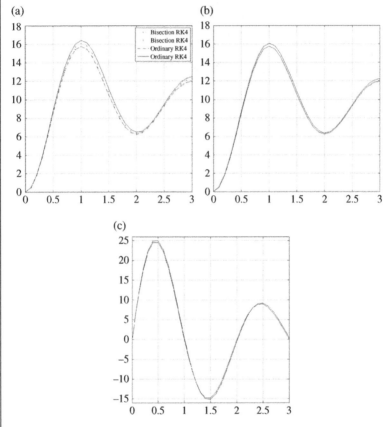

Figure 4.15 Ordinary and developed IRKM^{4th} on problem (a), ordinary and developed GIRKM^{4th} on X_1 (b) and X_2 (c) by considering the switching points.

Case Study 4.2   Forced Pendulum

In 1922, Georg Hamel was the first person who introduced a model of the forced pendulum equation with the rigorous mathematical study of the periodic solutions [55, 56]. The forced pendulum can be analyzed as a proper nonlinear dynamic system. Assume we have a heavy mass (*m*) on the end of a rigid body (rod) with length *L*. A small motor is placed on the other end of the rod to supply the torque, τ. The motor is driven in a sinusoidal fashion and the torque and frequency (ω) can be controlled. *g* is the gravity that acts downward and the angle of the pendulum (θ) is considered zero during the rest. A torque proportional to the angular velocity with a coefficient β is supplied by the friction in the motor and bearings.

By applying Newton's laws, the force balance can be achieved as follows:

$$\frac{d^2\theta}{dt^2} = -\frac{g}{L}\sin(\theta) - \frac{\beta}{mL}\frac{d\theta}{dt} + \frac{\tau}{mL}\sin(\omega t). \tag{4.129}$$

Figure 4.16 shows a simple model for the Forced Pendulum.

By simplification and numeric substitution of the problem coefficients with considering uncertainty [57], the main equation is turned into

$$\begin{cases} \dot{X}_1 = X_2 \\ \dot{X}_2 = A_0 X_2 - A_1 \sin(X_1) + A_2 \sin(t) \end{cases} \tag{4.130}$$

Assume that $A_0 \in [-0.3, -0.1]$, $A_1 \in [-1.5, -0.5]$, and $A_2 \in [1, 3]$.

The exact model of the forced pendulum is modulated by $A_0 = -0.2$, $A_1 = -1$, and $A_2 = 2$.

With the same initial value ($X_1(0) = X_2(0) = 0$), time interval, and step size in the previous problem, the solution can be obtained by the interval arithmetic as follows (Figure 4.17):

As can be observed from Figure 4.17, the exact model is placed in the interval model. The switching points X_1 and X_2 for this problem are given in Table 4.5.

Figure 4.16 Forced pendulum.

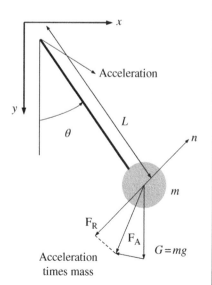

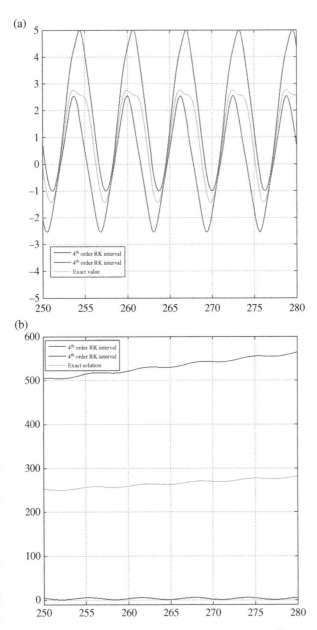

Figure 4.17 The focused result of the proposed GIRKM^{4th} on the forced pendulum on (a) X_1 and (b) X_2.

Table 4.5 Describing the switching points of the solution for the forced pendulum.

	Increasing	Decreasing
X_1	[0, 3] and [6, 9.4] and [12.4, 15.7]	[3, 6] and [9.4, 12.4] and [15,7–19.8]
X_2	[0, 5.4] and [9.2, 11.5] and [13.3–17.9]	[5.4, 9.2] and [11.5, 13.3], [17.9,18.8]

Table 4.5 indicates the switching points of the solution for the forced pendulum.

The nonlinear solution of the pendulum system with uncertainties X_1 and X_2 is shown in the figure above. As can be seen, all the orders of the GIRKM can almost follow the system solution.

Case Study 4.3   Lorentz Attractor

The Lorenz Attractor was initially introduced by Edward Lorenz to model the chaotic behavior exhibited in weather patterns. The original system consisted of 12 nonlinear differential equations. However, to simplify the model [58], Lorenz presented the following formulation for the Lorenz system:

$$\begin{cases} \dot{X}_1 = A_0(X_2 - X_1) \\ \dot{X}_2 = X_1(A_1 - X_3) - X_2 \\ \dot{X}_3 = X_1 X_2 - A_2 X_3 \end{cases} \tag{4.131}$$

where $A_0 \in [9, 11]$, $A_1 \in [26, 28]$, and $A_2 \in \left[\dfrac{6}{3}, \dfrac{10}{3}\right]$.

The variables X_1, X_2, and X_3 describe the physical quantities such as temperatures and flow velocities, while the uncertainty values, A_0, A_1, and A_2 describe the characteristics of the atmospheric system. Initial condition in this problem is selected as $[X_1, X_2, X_3] = [1,0,0]$. The results are shown in Figure 4.18, which illustrates the evolution using different GIRKM[4th].

Figure 4.18 shows that IRKM[1st] and IRKM[2nd] could not track the system response properly. The exact central solution for the Lorentz gives about a maximum of 18 in the time 0.4. But here except for the GIRKM[4th], the lower orders of GIRKMs exceed the value and even their lower interval could not include the central solution. From the result, if the order of the RK is getting higher, the interval result will envelop the possibility result tightly.

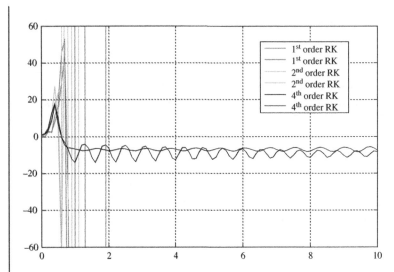

Figure 4.18 The result of the proposed GIRKM4th for the Lorentz Attractor.

Case Study 4.4   LQR Control

Consider a Linear Quadratic Regulator (LQR) with the performance index as follows:

$$J = \int_0^\infty \left(\frac{1}{2}x^T Q x + \frac{1}{2}u^T R u \right) dt, \tag{4.132}$$

where both Q and R should be positive definite.

By assuming the plant's dynamic as follows:

$$\dot{x} = Ax(t) + Bu(t), \tag{4.133}$$

And considering a state variable feedback regulator,

$$u = -\left(R^{-1}B'P() \right) \tag{4.134}$$

The matrix P should satisfy the reduced-matrix Riccati equation as follows [59]:

$$-\dot{P} = PAx + A^T Px + Qx - PBR^{-1}B^T Px, \tag{4.135}$$

which has to hold for all x [60].

Now consider the following Ricatti equation:

$$\dot{P} = \Gamma P^2(t) + 1; P(0) = 0. \tag{4.136}$$

where $\Gamma \in [-1.5, -0.5]$, $A = 1$, $B = -1$, $Q = 1$, and $H = 0$. The exact solution without uncertainty ($\Gamma = -1$) is $P(t) = \dfrac{e^{2t} - 1}{e^{2t} + 1}$.

By using the proposed GIRKM^{4th}, the solution will be obtained as in Figure 4.19.

(a)

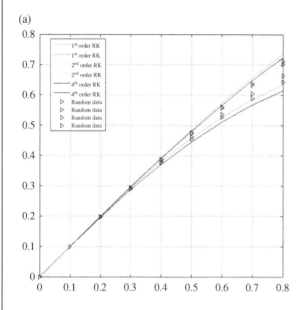

(b)

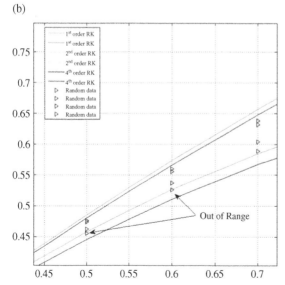

Figure 4.19 (a) The result of the proposed GIRKMs for the LQR system, (b) focused on the interval from 0.6 to 0.85.

In Figure 4.19, the exact method is also shown. Note that because in this problem, we have no switching points, the GIRKM is the same as IRKM. As can be concluded, the exact solution is a subset of the interval. For more analysis, three random Γ values are selected by the computer in the considered interval. The results from Figure 4.19b show that except IRKM^{1st}, the other methods (second- and fourth-order IRKMs) achieved good results.

The other point is that we should note that this method is efficient for IVP systems because of the RK methods' nature; since, for improving it to solve the BVP methods, we can improve it by some different techniques like the shooting method [61].

Therefore, IVP differential equations are studied in the presence of interval uncertainties by using a generalized Hukuhara-based Runge–Kutta method (GIRKM) from both theory and practical points of view. We have introduced new representations for Hukuhara differencing to improve the *IRKM* into a new interval method. The method is then utilized to solve some practical problems. For narrowing the interval in the proposed method, a bisection method can be also utilized to turn the main interval into two subintervals and after solving the equation, the unions of these two solutions have given the optimized interval. There are also some problems with the nonmonotonic equations. This problem is also solved by a technique based on critical points (switching points).

4.15 Interval Uncertainty Analyses based on Orthogonal Functions

One of the popular numerical methods for solving differential equations is the use of orthogonal spectral methods [62–66]. In recent years, the use of this method for numerical solutions and optimization of the cost function in optimal control problems has been considerably respected by researchers. Among the reasons for this attention are their low sensitivity, stability, high convergence rate, and the convenience of adding more basic functions.

There are always some uncertainties in solving real engineering problems. Unfortunately, the classical methods of orthogonal functions cannot solve such problems; therefore, if spectral functions can be transferred to the interval space, these problems will be resolved [67–69].

4.15.1 Interval ε-orthogonal

Two-vector $u, v \in \mathbb{IR}$ are called interval ε-orthogonal if, for the weight function, we have

$$([u], [v])_w = \varepsilon I_c \tag{4.137}$$

Theorem 4.5 The two vectors $u, v \in \mathbb{IR}$ based on the Kaucher extended IA, the weight function is orthogonal if they satisfy the upper bound and the lower bound is orthogonal.

Proof: By assuming the interval Kaucher multiplication of the two vectors

$$[v] = [[v_1], [v_2], ..., [v_n]]^T, [u] = [[u_1], [u_2], ..., [u_n]]^T$$
$$if \ [u], [v] \in \mathbb{IK} : [u].[v] = [\underline{uv}, \overline{uv}] \tag{4.138}$$
$$\forall \varepsilon \ll 0; [v] \bot [u] \Leftrightarrow \underline{uv}, \overline{uv} = \varepsilon I_c$$

As a result, by assuming the interval equality,

$$\underline{u_1}\underline{v_1} + \underline{u_2}\underline{v_2} + ... + \underline{u_n}\underline{v_n} = -\varepsilon$$
$$\overline{u}_1\overline{v}_1 + \overline{u}_2\overline{v}_2 + ... + \overline{u}_n\overline{v}_n = \varepsilon \tag{4.139}$$

Therefore, following the strategy of the Kaucher multiplication, for analyzing the interval orthogonality, we need only upper and lower bounds.

Note that the interval orthogonal is an approximated term, i.e., the interval orthogonal will be interval ε-orthogonal, which is different from that of a conventional orthogonal.

4.15.2 Interval Weierstrass's Theorem

Assuming that the interval function, *[F]* is continuous on the closed box *[B]*, there is a polynomial *[P_n]* for each $\varepsilon > 0$, where

$$\|[F(x) - P_n(x)]\| \in \varepsilon I_F \tag{4.140}$$

4.16 Interval Orthogonal Polynomials

Based on this generalization, the interval orthogonal polynomials having the proper properties are good approximations to solve the considered problem.

In the following, we introduce two types of the most well-known polynomials based on norm two (Legendre) and norm infinity (Chebyshev).

4.16.1 Legendre Polynomials

These polynomials are the solutions of the ordinary differential equation, called the Legendre differential equation, which is mathematically formulated as follows:

$$\frac{d}{dt}\left[(1-t^2)\frac{d}{dt}P_n(t)\right] + n(n+1)P_n(t) = 0 \tag{4.141}$$

Legendre polynomials (LPs) are perpendicular (orthogonal) to each other in the interval $[-1, 1]$ and with the weight function $w(t) = 1$. The recursive definition of the polynomial can be represented as follows [15, 83, 138]:

$$\begin{aligned} P_0(t) &= 1, \\ P_1(t) &= t, \\ P_{n+1}(t) &= \frac{n+2}{n}(t)P_n(t) - \frac{n}{n+1}P_{n-1}(t), \qquad n = 1, 2, \dots \end{aligned} \tag{4.142}$$

By changing the variable $\dfrac{\sin(n+1)\theta}{\sin\theta} = \sum_{l=0}^{n} P_l(\cos\theta)P_{n-l}(\cos\theta)$, the LPs can be defined by the following trigonometric functions:

$$P_n(t) = \cos(n\theta), \tag{4.143}$$

All of these polynomials are twofold orthogonal to the weight function, $w(t) = 1$. Given the least squares method and orthogonal definition,

$$P_n(t) \simeq c_0 P_0(t) + \sum_{j=1}^{k} c_j P_j(t) \tag{4.144}$$

where f is the main function and c_j is the Legendre expansion spectrum.

$$c_j = \frac{2j+1}{2}\int_0^\pi f(t)P_j(t)\mathrm{d}t, j = 0, 1, \dots, k \tag{4.145}$$

4.16.2 Chebyshev Polynomials

Chebyshev polynomials (CPs) are a sequence of orthogonal polynomials. The names of these polynomials were derived from the name of the Russian mathematician, Pafnuty Lvovich Chebyshev, who introduced them first in 1854.

Chebyshev functions are obtained from the solution of the following equations [70–72]:

$$\begin{aligned} (1-t^2)T_n'' - tT_n' + n^2 T_n &= 0, \\ (1-t^2)T_n'' - 3tT_n' + n(n+2)T_n &= 0 \end{aligned} \tag{4.146}$$

CPs are also orthogonal in the above sense, in the interval $[-1, 1]$ and terms of weight function $w(t) = \dfrac{1}{\sqrt{1-t^2}}$. This polynomial can be recursively calculated by the following formula:

$$
\begin{aligned}
&T_0(t) = 1, \\
&T_1(t) = x, \\
&T_{n+1}(T) = 2tT_n(t) - T_{n-1}(t), n = 1,2,3,\ldots \\
&T_n(t) = \cos(n\theta),
\end{aligned}
\tag{4.147}
$$

By changing the variable $\theta = accros(x)\epsilon[0, \pi]$, CPs are shown as the following trigonometric equation:

$$
T_n(t) = \cos(n\theta),
\tag{4.148}
$$

All of these polynomials are orthogonal with each other by the weight function, $\omega(t) = 1$, i.e.,

$$
\int_{-1}^{1} \omega(t)T_m(t)T_n(t)\mathrm{d}t = \int_{j=0}^{\pi} \cos m\theta \ \cos n\theta\mathrm{d}\theta = \begin{cases} \pi, r = s \neq 0 \\ \dfrac{\pi}{2}, r = s \neq 0 \\ 0, r = s \neq 0 \end{cases}
\tag{4.149}
$$

The final formula to find the CP can be summarized as follows:

$$
t = [\underline{t},\overline{t}]
\tag{4.150}
$$

where f is the main function and T_j is the exponential spectra of the Chebyshev extension. Some important features of these polynomials are given below:

1) $T_n(x) \mid \ \leq 1 for - 1 \leq x \leq 1 \forall n.$
2) The lowest and highest values for $\mid T_n \mid$ in the interval $x \in [-1, 1]$ for $n+1$ points are -1 and 1.
3) $\mid T_n \mid$ has n number of specified zero in the interval $x \in [-1, 1]$.

For more understanding, see Chapter 3 of [73].

The error resulting from the truncation error between the Chebyshev function and the main function is

$$
e_n(t) = \mid f(t) - P_n(t) \mid \ = \dfrac{2^{-n}}{(n+1)!} \left\| f^{(n+1)} \right\|_{\infty}
\tag{4.151}
$$

If n is large enough, $C_n(t)$ can be discarded. The restricted Chebyshev series have high precision approximation compared to other approximations.

4.16.3 Interval Orthogonal Functions

We can extend the main function given by orthogonal functions (such as Chebyshev functions and Legendre functions) and then use IA to generate the orthogonal functions of the interval orthogonal functions. By replacing the value of the interval value $[x]$ with x in the equations, the orthogonal functions are obtained as follows:

$$[f]([x]) = \frac{1}{2}f_0 + \sum_{i=1}^{h} f_i \phi_i([x]), \tag{4.152}$$

where $\phi_i([x])$ represents the core of the orthogonal functions.

Each orthogonal function is in a certain range; in this case, the core of the Chebyshev functions $[-1, 1]$ is approximated in the interval.

In general, the use of trigonometric definition in orthogonal functions leads to better results, since it is easier to solve and has fewer errors briefly [74–76], the trigonometric definition can be described as

$$[f]([x]) = \frac{1}{2}c_0 + \sum_{i=1}^{h} c_i \times g(\cos(i[t])), \tag{4.153}$$

where $g(\cos(i[t]))$ indicates that it is a cosine-dependent value, e.g., for the Chebyshev functions $[g]([\cos(it)]) = [\cos](it)$, the values are dependent on the cosine value. Therefore, the point will be reached when Chebyshev functions, $-1 \leq [g]$ $([\cos(it)]) \leq 1$.

Therefore, by considering the interval limitations,

$$[f]([x]) = \frac{1}{2}c_0 + I_c \times \sum_{i=1}^{h} |c_i| \tag{4.154}$$

where c_i is the expansion coefficient of the orthogonal functions.

As it is known, the definition of the orthogonal function arrays in the interval is defined based on the centered definition. Note that all the orthogonal functions can be generalized using forward, backward, or centered definitions mentioned before.

In the case of using orthogonal functions in the range $[-1,1]$, if the lower interval $t_0 \neq -1$, or the high interval $t_f \neq 1$, the following transformation should be applied:

$$\tau = \frac{t_f - t_0}{2}t + \frac{t_f + t_0}{2}, \tag{4.155}$$

In the following, the extension of the two discussed orthogonal functions is mentioned. Let us assume $y = 2e^{[x]} + \sin([x]) + \tan^{-1}([x])$ in $[x] = [-1, 1]$.

The extension of this function is shown by the orthogonal functions of the Legendre and Chebyshev functions as follows:

$$[f_{c_5}]([x]) = 2.5321 + 3.9692[\cos]([\theta]) + 0.543[\cos](2[\theta])$$
$$+ 0.0021[\cos](3[\theta]) + 0.0109[\cos](4[\theta]) + 0.007[\cos](5[\theta])$$
$$= [-7.0579, 7.0579],$$

$$(4.156)$$

$$[f_{P_5}]([x]) = 2.3504 + 3.9670(\cos([t])) + 0.7156g([\cos](2[\theta]))$$
$$- 0.0021g([\cos](3[\theta])) + 0.0199g([\cos](4[\theta])) \qquad (4.157)$$
$$+ 0.0138g([\cos](5[\theta])) = [-7.0551, 7.0551],$$

4.17 Piecewise Extension of the Interval Orthogonal Functions

Several problems in nature can be modeled as continuous piecewise functions. For example, in cases where a scale change causes costs to be raised or reduced, or in cases where a controller is required for switching (on/off), we need to work with these types of functions.

Typically, systems that are inherently switched are modeled using the theory of moderation with a non-linear nonswitching system. One of the methods of studying nonlinear systems is the use of continuous piecewise functions.

Using the Chebyshev interval method is much better to reduce the extra range (wrapping effect). In this section, a modified piecewise method is proposed for the interval Chebyshev method to apply to switching problems.

The advantage of this method is the function's faster operation and lower error estimation. The reason for the increase in speed, in this case, is that to achieve a precise solution of a function, there is a need for higher order (n) of the quotient polynomials, which in turn would increase the cost of the method, but, to [77], the use of piecewise functions with even lower order has a better accuracy and speed.

In this method, the time interval is first divided into several subintervals with lower order, and then each subspace is transmitted using the interval Chebyshev method, and given that the CP in the interval $[-1, 1]$ is solved.

For instance, if the interval is divided into two subintervals, two intervals $[-1, 1]$ and $[0\ 1]$ that are solved separately and based on the transmitted Chebyshev functions will be generated. By recursive extending of this subject and assuming a real interval $[X] = [\underline{x}, \overline{x}]$:

$$\text{for } x = 1 : [X_1] = \left[\underline{x}, \underline{x} + \frac{1}{k}x_w\right],$$
$$\text{for } x > 1 : [X_{k+1}] = \left[\overline{x}_k, \overline{x}_k + \frac{1}{k}x_w\right], k = 1, 2, ..., n$$

$$(4.158)$$

where k describes the number of sub-sections.

To solve the system under the divided intervals based on the above method, a conversion is required on the Chebyshev function from the interval $[-1, 1]$ into each of the following subintervals. So, assuming having a subinterval $t = [\underline{t}, \overline{t}]$ in which $t \in [-1, 1]$, the converted function is obtained as follows:

$$\tau = \frac{\overline{t} + \underline{t}}{2} + \frac{\overline{t} - \underline{t}}{2}[-1, 1], \tag{4.159}$$

where τ represents the transposed range.

One of the advantages of this method is that we use a lower order of Chebyshev functions to approximate the intervals.

Numerical example: Consider the function $F([X]) = \arctan([X])$ in the interval $[X] \in [-1, 1]$.

For the approximation of the function, the interval Chebyshev function of order 3 and piecewise interval Chebyshev function of order 10 for the equations described above are employed.

By dividing the interval into 5 and 10 subintervals and applying the Chebyshev to the function, the results are obtained as follows (Note that the horizontal lines are only for showing the endpoint of the previous function and the initial point of the next function.)

As can be observed from Figure 4.20, the use of the interval piecewise Chebyshev function greatly increases the compression of the interval method, but it should be

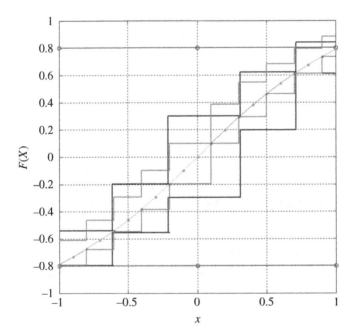

Figure 4.20 Estimation of the function $F([X]) = \arctan([X])$ in the interval $[X] \in [-1, 1]$ using the interval Chebyshev (blue lines) function and the interval piecewise Chebyshev function of order 5 for 5 subintervals (black lines) and 10 subintervals (green lines).

noted that due to the reduction of the accuracy of the CPs in this case, the possibility of decreasing the accuracy of the inclusion, there will also be an exact range of uncertainties.

4.18 Conclusion

In this chapter, the IA and its variants, such as the classical (Minkowski) methods, model-based methods, and methods based on differentials and derivatives of Hukhara, were introduced. The advantages and applications of each method in solving indeterminate problems were also studied. In the following, with the introduction of three types of representations, the definitions were rewritten. After introducing different types of illustrations, important definitions such as interval functions and derivation that were used to solve the optimal control systems were introduced. In the next section, unconventional dynamic systems were introduced based on interval uncertainties, and a new multimethod strategy based on the Runge–Kutta strategy was introduced to solve these types of problems. In the chapter, with the introduction of orthogonal spectral functions and related topics, the necessary preconditions for modeling and analyzing the optimal control problems in different systems were provided. Here, the general properties of the spectral theory were also referred to. Two popular types of orthogonal polynomials, which were well-known for accuracy and speed (LPs and CPs), were defined. Finally, a piecewise version of the Chebyshev method was introduced as a way to speed up and reduce the wrapping effect error.

References

1 Alefeld, G., and G. Mayer, *Interval analysis: theory and applications.* Journal of Computational and Applied Mathematics, 2000. **121**(1): p. 421–464.

2 Li, P., W. Xu, and S. Li-Wei, *Power pattern tolerance analysis of radome with the material property error based on interval arithmetic.* IEEE Antennas and Wireless Propagation Letters, 2017. **16**: p. 1321–1324.

3 Dawood, H., *Theories of interval arithmetic: mathematical foundations and applications.* 2011, LAP Lambert Academic Publishing.

4 Li, T., et al., *Accuracy analysis and form-finding design of uncertain mesh reflectors based on interval force density method.* Proceedings of the Institution of Mechanical Engineers, Part G: Journal of Aerospace Engineering, 2017. **231**(11): p. 2163–2173.

5 Pérez-Galván, C., and I. D. L. Bogle, *Global optimisation for dynamic systems using interval analysis.* Computers & Chemical Engineering, 2017. **107**: p. 343–56.

6 Stefanini, L., and B. Bede, *Generalized Hukuhara differentiability of interval-valued functions and interval differential equations.* Nonlinear Analysis: Theory, Methods & Applications, 2009. **71**(3): p. 1311–1328.

7 Chui, C. K., and G. Chen, *Kalman filtering for interval systems*, in *Kalman Filtering*. 2017, Springer. p. 151–170.

8 Luo, J., L. Shi, and Y. Ni, *Uncertain power flow analysis based on evidence theory and affine arithmetic.* IEEE Transactions on Power Systems, 2017. **33**: p. 1113.

9 Giusti, A., and M. Althoff. *Efficient computation of interval-arithmetic-based robust controllers for rigid robots.* in *Robotic Computing (IRC), IEEE International Conference on* 10–12 April 2017, Taichung, Taiwan, IEEE.

10 Guo, S. X., and Y. Li, *Robust reliability method and reliability-based performance optimization for non-fragile robust control design of dynamic system with bounded parametric uncertainties.* Optimal Control Applications and Methods, 2017. **38**(2): p. 279–292.

11 Viegas, C., et al., *Performance analysis and design of parallel kinematic machines using interval analysis.* Mechanism and Machine Theory, 2017. **115**: p. 218–236.

12 Li, P., et al., *Far-field pattern tolerance analysis of the antenna-radome system with the material thickness error: an interval arithmetic approach.* IEEE Transactions on Antennas and Propagation, 2017. **65**(4): p. 1934–1946.

13 Jeyasenthil, R. and P. Nataraj, *An interval-consistency-based hybrid optimization algorithm for automatic loop shaping in quantitative feedback theory design.* Journal of Vibration and Control, 2017. **23**(3): p. 414–431.

14 The Institute of Electrical and Electronics Engineers, Inc. *IEEE Approved Draft Standard for Interval Arithmetic (Simplified).* IEEE P1788.1/D9.9, June 2017, 2017: p. 1–36.

15 Revol, N. *Introduction to the IEEE 1788-2015 standard for interval arithmetic.* in *International Workshop on Numerical Software Verification.* 2017. Springer.

16 Kaucher, E., *Interval analysis in the extended interval space IR*, in *Fundamentals of Numerical Computation (Computer-Oriented Numerical Analysis).* 1980, Springer. p. 33–49.

17 Gardeñes, E., H. Mielgo, and A. Trepat, *Modal intervals: reason and ground semantics*, in *Interval Mathematics 1985.* 1986, Springer. p. 27–35.

18 Elskhawy, A., K. Ismail, and M. Zohdy. Modal interval floating point unit with decorations. in *16th GAMM-IMACS International Symposium on Scientific Computing, Computer Arithmetic and Validated Numerics.* 2014. University of Würzburg Germany

19 Hayes, N. T., Introduction to modal intervals. available at grouper. IEEE. org/groups/1788/Material/Hayes ModalInterval. pdf, 1788.

20 Hukuhara, M., *Integration des applications mesurables dont la valeur est un compact convexe.* Fako de l'Funkcialaj Ekvacioj Japana Matematika Societo, 1967. **10**: p. 205–223.

21 Malinowski, M. T., *Interval Cauchy problem with a second type Hukuhara derivative.* Information Sciences, 2012. **213**: p. 94–105.

22 Lupulescu, V., *Hukuhara differentiability of interval-valued functions and interval differential equations on time scales.* Information Sciences, 2013. **248**: p. 50–67.

23 Gomes, L. T., and L. C. Barros, *A note on the generalized difference and the generalized differentiability.* Fuzzy Sets and Systems, 2015. **280**: p. 142–145.

24 Stefanini, L., *A generalization of Hukuhara difference and division for interval and fuzzy arithmetic.* Fuzzy Sets and Systems, 2010. **161**(11): p. 1564–1584.

25 Bede, B., and L. Stefanini, *Generalized differentiability of fuzzy-valued functions.* Fuzzy Sets and Systems, 2013. **230**: p. 119–141.

26 Hüllermeier, E., *An approach to modelling and simulation of uncertain dynamical systems.* International Journal of Uncertainty, Fuzziness and Knowledge-Based Systems, 1997. **5**(2): p. 117–137.

27 Moore, R. E., R. B. Kearfott, and M. J. Cloud, *Introduction to interval analysis.* 2009, SIAM.

28 Yang, X., J. Cao, and J. Liang, *Exponential synchronization of memristive neural networks with delays: interval matrix method.* IEEE Transactions on Neural Networks and Learning Systems, 2017. **28**: 1878.

29 Qiao, J., and B. Q. Hu, *On interval additive generators of interval overlap functions and interval grouping functions.* Fuzzy Sets and Systems, 2017. **323**: 19.

30 Ma, H., S. Xu, and Y. Liang, *Global optimization of fuel consumption in J 2 rendezvous using interval analysis.* Advances in Space Research, 2017. **59**(6): p. 1577–1598.

31 Breuel, T. M., *On the use of interval arithmetic in geometric branch and bound algorithms.* Pattern Recognition Letters, 2003. **24**(9): p. 1375–1384.

32 Jaulin, L., *Applied interval analysis: with examples in parameter and state estimation, robust control and robotics.* Vol. 1. 2001: Springer Science & Business Media.

33 Saraev, P. V., *Numerical methods of interval analysis in learning neural network.* Automation and Remote Control, 2012. **73**(11): p. 1865–1876.

34 Rihm, R., *On a class of enclosure methods for initial value problems.* Computing, 1994. **53**(3–4): p. 369–377.

35 Nedialkov, N. S., *Computing rigorous bounds on the solution of an initial value problem for an ordinary differential equation.* 1999, Citeseer.

36 Stauning, O., and K. Madsen, *Automatic validation of numerical solutions.* 1997, Technical University of DenmarkDanmarks Tekniske Universitet, Department of Informatics and Mathematical ModelingInstitut for Informatik og Matematisk Modellering.

37 Madadi, A., N. Razmjooy, and M. Ramezani, *Robust control of power system stabilizer using world cup optimization algorithm.* International Journal of Information, Security and Systems Management, 2016. **5**(1): p. 519–526.

38 Schutze, O., et al., *Using the averaged Hausdorff distance as a performance measure in evolutionary multiobjective optimization.* IEEE Transactions on Evolutionary Computation, 2012. **16**(4): p. 504–522.

39 Moran, P. A. *Additive functions of intervals and Hausdorff measure.* in *Mathematical Proceedings of the Cambridge Philosophical Society.* 1946, Cambridge Univ Press.

40 Ayer, E., and R. Strichartz, *Exact Hausdorff measure and intervals of maximum density for Cantor sets.* Transactions of the American Mathematical Society, 1999. **351**(9): p. 3725–3741.

41 Hansen, E. R., and R. I. Greenberg, *An interval Newton method.* Applied Mathematics and Computation, 1983. **12**(2–3): p. 89–98.

42 Mori, H. and A. Yuihara. *Calculation of multiple power flow solutions with the Krawczyk method.* in *1999 IEEE International Symposium on Circuits and Systems (ISCAS).* 30 May 1999–2 June 1999, Orlando, FL, USA, IEEE.

43 Ebrahimi, N., et al., *Tolerance analysis of the reflectarray antenna through Minkowski-based interval analysis.* in *Antennas and Propagation (EUCAP), 2017 11th European Conference on.* 2017. IEEE.

44 Egiziano, L., et al., *Robust design of electromagnetic systems based on interval taylor extension applied to a multiquadric performance function.* IEEE Transactions on Magnetics, 2008. **44**(6): p. 1134–1137.

45 Edmonson, W., and G. Melquiond. *IEEE interval standard working group – P1788: current status.* in *2009 19th IEEE Symposium on Computer Arithmetic.* 2009, Online resources for the IEEE Interval Standard Working Group.

46 Dorf, R. C. and R. H. Bishop, *Modern control systems.* 2011, Pearson.

47 Akram, S., and Q. U. Ann, *Newton raphson method.* International Journal of Scientific & Engineering Research, 2015. **6**(7): p. 1748–1752.

48 Li, Q., Z. Qiu, and X. Zhang, *Eigenvalue analysis of structures with interval parameters using the second-order Taylor series expansion and the DCA for QB.* Applied Mathematical Modelling, 2017. **49**: p. 680.

49 Liao, X., et al., *Interval method for uncertain power flow analysis based on Taylor inclusion function.* IET Generation, Transmission & Distribution, 2017. **11**(5): p. 1270–1278.

50 Wu, J., et al., *Uncertain analysis of vehicle handling using interval method.* International Journal of Vehicle Design, 2011. **56**(1–4): p. 81–105.

51 Ravat, D., *Analysis of the Euler method and its applicability in environmental magnetic investigations.* Journal of Environmental and Engineering Geophysics, 1996. **1**(3): p. 229–238.

52 Chalco-Cano, Y. and H. Román-Flores, *On new solutions of fuzzy differential equations.* Chaos, Solitons & Fractals, 2008. **38**(1): p. 112–119.

53 Wu, J., et al., *A new uncertain analysis method and its application in vehicle dynamics.* Mechanical Systems and Signal Processing, 2015. **50**: p. 659–675.

54 Nedialkov, N. S., K. R. Jackson, and G. F. Corliss, *Validated solutions of initial value problems for ordinary differential equations.* Applied Mathematics and Computation, 1999. **105**(1): p. 21–68.

55 Girg, P. and F. Roca, *On the range of certain pendulum-type equations.* Journal of Mathematical Analysis and Applications, 2000. **249**(2): p. 445–462.

56 Mawhin, J., *Global results for the forced pendulum equation.* Handbook of Differential Equations: Ordinary Differential Equations, 2000. **1**: p. 533–589.

57 Hu, J.-W., and H.-M. Tang, *Numerical methods for differential equations.* City University, Hong Kong, 2003.

58 Azar, A. T., and S. Vaidyanathan, *Chaos modeling and control systems design.* 2015, Springer.

59 Razmjooy, N., M. Ramezani, and E. Nazari, *Using LQG/LTR optimal control method for car suspension system.* SCRO Research Annual Report, 2015. **3**: p. 1–8.

60 Razmjooy, N., et al., *Comparison of lqr and pole placement design controllers for controlling the inverted pendulum.* Journal of World's Electrical Engineering and Technology, 2014. **2322**: p. 5114.

61 Pries, J., and H. Hofmann, *Steady-state algorithms for nonlinear time-periodic magnetic diffusion problems using diagonally implicit runge–kutta methods.* IEEE Transactions on Magnetics, 2015. **51**(4): p. 1–12.

62 Bhrawy, A., et al., *Solving fractional optimal control problems within a Chebyshev–Legendre operational technique.* International Journal of Control, 2017. **90**(6): p. 1230–1244.

63 Chen, C., et al., *Blind forensics of successive geometric transformations in digital images using spectral method: theory and applications.* IEEE Transactions on Image Processing, 2017. **26**(6): p. 2811–2824.

64 Coll-Font, J., B. Erem, and D. H. Brooks, *A potential-based inverse spectral method to non-invasively localize discordant distributions of alternans on the heart from the ECG.* IEEE Transactions on Biomedical Engineering, 2017. **65**: 1554.

65 Ezz-Eldien, S., et al., *A numerical approach based on Legendre orthonormal polynomials for numerical solutions of fractional optimal control problems.* Journal of Vibration and Control, 2017. **23**(1): p. 16–30.

66 Guo, S., L. Mei, and Y. Li, *An efficient Galerkin spectral method for two-dimensional fractional nonlinear reaction–diffusion-wave equation.* Computers & Mathematics with Applications, 2017. **74**: p. 2449.

67 Li, C., et al., *Sparse regression Chebyshev polynomial interval method for nonlinear dynamic systems under uncertainty.* Applied Mathematical Modelling, 2017. **51**: p. 505.

68 Wang, Z.-q., and J. Mu, *A multiple interval chebyshev-gauss-lobatto collocation method for ordinary differential equations.* Numerical Mathematics: Theory, Methods and Applications, 2016. **9**(4): p. 619–639.

69 Wu, J., et al., *Interval uncertain method for multibody mechanical systems using Chebyshev inclusion functions*. International Journal for Numerical Methods in Engineering, 2013. **95**(7): p. 608–630.

70 Christiansen, J. S., B. Simon, and M. Zinchenko, *Asymptotics of Chebyshev polynomials, I: subsets of {\mathbb {R}}*. Inventiones Mathematicae, 2017. **208**(1): p. 217–245.

71 Ahmadian, A., S. Salahshour, and C. S. Chan, *Fractional differential systems: a fuzzy solution based on operational matrix of shifted Chebyshev polynomials and its applications*. IEEE Transactions on Fuzzy Systems, 2017. **25**(1): p. 218–236.

72 Protasov, V. Y., and R. M. Jungers, *Analysing the stability of linear systems via exponential Chebyshev polynomials*. IEEE Transactions on Automatic Control, 2016. **61**(3): p. 795–798.

73 Rivlin, T. J., *An introduction to the approximation of functions*. 2003, Courier Corporation.

74 Ghosh, S., and K. N. Chaudhury. Fast and high-quality bilateral filtering using Gauss-Chebyshev approximation. in *Signal Processing and Communications (SPCOM), 2016 International Conference on* 12–15 June 2016, Bangalore, India, IEEE.

75 Wu, J., *Uncertainty analysis and optimization by using the orthogonal polynomials*. Doctoral dissertation. 2015, Faculty of Engineering and Information Technology, University of Technology, Sydney.

76 Bhatnagar, V., R. Majhi, and S. T. Devi, *Development of an efficient prediction model based on a nature-inspired technique for new products: a case of industries from the manufacturing sector*, in *Handbook of Research on Modeling, Analysis, and Application of Nature-Inspired Metaheuristic Algorithms*. 2018, IGI Global. p. 160–182.

77 Ma, H., E. A. Butcher, and E. Bueler, *Chebyshev expansion of linear and piecewise linear dynamic systems with time delay and periodic coefficients under control excitations*. Journal of Dynamic Systems, Measurement, and Control, 2003. **125**(2): p. 236–243.

5

Stability and Controllability Based on Interval Analysis

5.1 Introduction

Control engineering utilizes feedback loops to regulate and manipulate the behavior of real systems. The objective is to ensure the system functions as intended and achieves its goals. Generally, to control a real system, the following steps must be considered:

Step 1: System identification. It is important to determine whether the system model is structured or unstructured. If the system is structured, check whether the model is definitive or if it contains uncertain parameters. Additionally, if there is uncertainty, identify the type of uncertainty (interval, random, chaotic, etc.).

Step 2: Analyze system stability and controllability. This step is crucial because if the system is not controllable (or at least stabilizable), any further operations will be in vain and the system will not be effectively controlled. Therefore, ensuring system controllability is a prerequisite for proceeding with the subsequent steps.

Step 3: Select the type of controller. The choice of controller type should align with the specific characteristics of the system.

To enhance understanding of interval optimal control and its distinctions from other strategies, a brief explanation of various control strategies is provided below.

5.1.1 Classical Control Theory

The classical control theory, often referred to as the oldest form of control, primarily focuses on the analysis and design of single-input single-output (SISO) systems. This theory is typically based on Laplace's theory, which enables system analysis using transfer functions. Figure 5.1 presents a simple block diagram depicting a closed-loop control system.

Interval Analysis: Application in the Optimal Control Problems, First Edition. Navid Razmjooy.
Published 2024 by John Wiley & Sons, Inc.

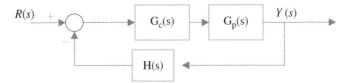

Figure 5.1 A simple structure of a closed loop classic control.

where $G_p(s)$, $G_c(s)$, and $H(s)$ are the process block, the controller block, and the sensor, respectively, and

$$\frac{Y(s)}{R(s)} = \frac{G(s)}{1 + G(s)H(s)}, \tag{5.1}$$

$$G(s) = G_p(s)G_c(s) \tag{5.2}$$

The analysis and control techniques in classical control theory primarily rely on graphical methods, making them relatively easy to solve manually. However, with the progress of computer and digital science, the utilization of these methods has diminished.

Classical control theory is primarily applicable to SISO systems. Extending its application to multivariate problems is challenging, even though most real-world systems are multi-input multi-output (MIMO).

5.1.2 Advanced Modern Control Systems Theory

As stated in the previous section, the focus of classical control systems theory is on SISO systems. To overcome this problem, an advanced modern control systems theory was introduced. This theory models MIMO based on the set of first-order differential equations. Here, the model of an linear time-invariant (LTI) system is expressed using state space equations as follows:

$$\begin{aligned} \dot{x}(t) &= Ax(t) + Bu(t) \\ y(t) &= Cx(t) + Du(t) \end{aligned} \tag{5.3}$$

where $A \in \mathbb{R}^{n \times n}$ represents the state matrix, $B \in \mathbb{R}^{n \times r}$ is the input matrix, and $C \in \mathbb{R}^{m \times n}$ and $D \in \mathbb{R}^{m \times r}$ are the output and the state transition matrices, respectively.

In general, for nonlinear systems:

$$\begin{aligned} \dot{x}(t) &= f(x(t), u(t), t) \\ y(t) &= g(x(t), u(t), t) \end{aligned} \tag{5.4}$$

The general configuration of the advanced control is shown in Figure 5.2.

Figure 5.2 A simple structure of an advanced control.

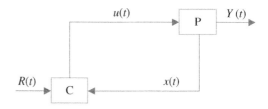

Among the distinguishing features of advanced modern control systems is the generation of the control signal, $\underline{u}(t)$, based on the system's state variables and the input signal, $R(t)$. This approach utilizes matrix theory, providing designers with greater flexibility even for complex systems. Table 5.1 presents a comprehensive comparison between classical control theory and advanced modern control systems theory.

5.1.3 Optimal Control

Optimal control is a critical component of advanced modern control theory. Optimal control for MIMO systems offers greater convenience by aiming to optimize the performance index, such as minimizing tracking time or reducing energy

Table 5.1 Comparison between the classical control theory and advanced modern control theory.

Classic control	Advanced control
Based on the transfer function	Based on state space
Based on the input and output relation or the transfer function.	System equations are described by n first-order differential equation
Incapable of solving nonlinear and variable systems.	Applicable to all types of linear, nonlinear, variable, SISO, and MIMO systems.
Initial conditions are considered zero	Initial conditions are: preserved.
It is a frequency domain approach	It is a time-domain approach
The transfer function of the system is unique	The state space solution is not unique
Input and output variables must be measurable.	It is not necessary that the variables of the state of the system are physical and even do not need to be measurable or visible.
There is no information about the internal mode of the system	The ability to design optimal and adaptive systems

consumption. There are two fundamental issues with classical control methods and advanced modern control methods:

1) These methods often overlook the primary objective and instead focus on global design, neglecting the value of optimal control.
2) Many practical processes encounter limitations, such as high costs, expensive control signals, wide ranges of control signals, or difficulties in generating the desired signals.

In such cases, optimal control can overcome these limitations. There are three main reasons for adopting optimal control theory over classical control theory and advanced modern control theory:

1) It can be applied to MIMO systems, which is not possible with classical control theory.
2) It allows for incorporating constraints during the design process.
3) It provides an optimal solution by nature.

5.1.4 Robust Control

As mentioned earlier, the goal of optimal control is to optimize a performance index in the time domain while considering physical constraints. On the other hand, robust control aims to enhance the stability and control quality of uncertain systems, accounting for uncertainties in the system, feedback sensors, and operations.

In optimal control, assuming a complete and controllable model, the control is optimized based on a specific objective. If the system is not complete according to the given definition, optimal control will not yield an optimal result. In contrast, robust control assumes that the system is not fully known or complete. For instance, suppose that the model parameters are incomplete but known within a certain range. Figure 5.3 shows the configuration of the robust control.

The primary objective of robust control theory is to manage systems that have uncertainties, particularly unstructured uncertainties. The ultimate goal is to achieve stability and satisfactory performance for each $p \in P$. Hence, one limitation of optimal control theory compared to robust control is the lack of a robustness guarantee for the system.

Although the fundamental purposes of optimal control and robust control differ, if it is possible to incorporate the advantage of robustness into the optimal control system,

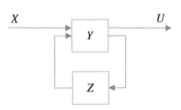

Figure 5.3 Configuration of the robust control.

Figure 5.4 Block diagram of adaptive control.

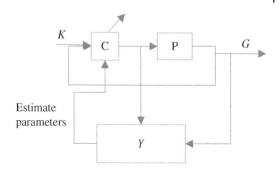

it will not only enhance realism at a minimal cost but also provide robustness. However, it is important to note that by combining system robustness with optimality, the resulting optimal controller will not be absolute and will become a suboptimal approach.

5.1.5 Adaptive Control Theory

Adaptive control is a more robust method than robust control! When the system changes based on time variations or changes in the system are too large, adaptive control will respond more appropriately.

So if we could not achieve a low-cost and robust method in a system, we have to use adaptive control. By assuming $\dot{x} = f(x, a, b)$ and $a, b \in S$, adaptive control is appropriate to estimate these parameters and then generate the control rule. Figure 5.4 shows the block diagram of the adaptive control.

In robust control, considering a system model $\dot{x} = f(x) + \delta f$, the objective is to achieve good system performance within a specified range without modifying the control policy, typically obtained through the H_∞ approach (worst-case scenario).

This approach differs from adaptive control, where the control policy is continually updated through adjustments. The fundamental principles underlying these methods are completely distinct. The purpose of presenting this section is solely to provide clarity to readers regarding the distinctions among these controllers.

5.2 Interval Stability and Controllability

Optimal control theory, in general, serves as a complementary approach to enhance system efficiency. However, it is important to note that if the system lacks stability and controllability, the optimal control problem becomes irrelevant. This assumption is always considered in conventional continuous optimal control problems like linear quadratic regulator (LQR) and linear quadratic tracking (LQT) [1].

Given that the primary objective is to analyze and determine the confidence interval for optimal control in the presence of uncertainty within a defined space, there is a possibility of encountering uncontrollability and instability when applying optimal control methods.

Hence, in this section, we investigate the stability and controllability of the interval system. The goal is to establish stability and controllability within a confidence interval so that a suitable control strategy can be designed. Put simply, if these two characteristics are absent in the system, optimal control becomes meaningless. The mathematical model of a dynamical system with finite interval uncertainty can be defined as follows.

$$\dot{x}(t) = [A]x(t) + [B]u(t),$$
$$x(t_0) = X_0, x(t_f) = X_f \quad\quad (5.5)$$
$$y(t) = Cx(t)$$

where $x(t) \in \mathbb{R}^n$ and $u(t) \in \mathbb{R}^m$ are the path and the control parameters, respectively. $y(t) \in \mathbb{R}^l$ is the output ($l < n$) and $[A] \in \mathbb{IR}^{n \times n}$ and $[B] \in \mathbb{IR}^{n \times m}$ are the interval matrices with the components $[a_{ij}] = [\underline{a}_{ij}, \overline{a}_{ij}]$ and $[b_{ij}] = [\underline{b}_{ij}, \overline{b}_{ij}]$ belonging to the interval set $\mathbb{IR} = \{[\underline{r}, \overline{r}] \mid \underline{r}, \overline{r} \in \mathbb{R}\}$ and the output matrix $C \in \mathbb{IR}^{l \times n}$. Considering a noninterval amount,

$$\dot{x}(t) = Ax(t) + Bu(t),$$
$$x(t_0) = X_0, x(t_f) = X_f \quad\quad (5.6)$$

where $A \in \mathbb{R}^{n \times n}$ and $B \in \mathbb{R}^{n \times m}$ are included in the $A \in [A], B \in [B]$.

5.3 Interval Stability

We first assume that the system Ω is defined with no uncertainty. The input is applied as a signal to the system. The vector of total inputs and the outputs at time t is defined by $u(t)$ and $y(t)$. It is assumed that the analyzed system is continuous. When Ω is linear and time-independent, its transfer function is defined as follows.

$$Y(s) = G(s)U(s) \quad\quad (5.7)$$

where s is the Laplace transform and $G(s)$ is the transfer function.

A common definition for the linear and time-independent systems is the state space definition:

$$\Omega : \begin{cases} \dot{x}(t) = Ax(t) + Bu(t) \\ y(t) = Cx(t) + Du(t) \end{cases} \tag{5.8}$$

where x is the state vector; $\dot{x}$ is the first derivative of x in time; and A, B, and C are state, control, and observe matrices, respectively.

Therefore, the system Ω is defined using a set of dynamical differential equations. Laplace transform definition of a state space for zero initial condition is

$$\begin{cases} s\dot{x}(s) = Ax(s) + Bu(s) \\ y(s) = Cx(s) + Du(s) \end{cases} \tag{5.9}$$

By eliminating the equation above, we have $y(s) = C(sI_n - A)^{-1}Bu(s)$, where I_n is the $n \times n$ identity matrix.

The system Ω is stable if the output is bounded for a bounded input signal, that is, by assuming

$$if: \forall t \geq 0, \exists u_m > 0 :| u(t) | \leq u_m < +\infty \tag{5.10}$$

to conclude

$$if: \forall t \geq 0, \exists y_m > 0 :| y(t) | = y_m < +\infty \tag{5.11}$$

5.4 Characteristic Polynomial

If $t \geq 0$ and $u(t) = 0$, the state vector is converted to the following definition:

$$x \times = e^{At}x(0) \tag{5.12}$$

Since the matrix arrays of e^{At} are linear compositions of the form $e^{\lambda_i t}$, in which, $\lambda_i t$ are the values of A, the system Ω will be stable if and only if all the special values of the real part are negative; in other words, $Re(\lambda_i) < 0$, $i = 1, 2, ..., n$.

Now, the eigenvalue of A is the root of the characteristic polynomial A, which is defined as follows:

$$P(s) = det(sI_n - A) \tag{5.13}$$

These polynomials can be formulated as follows:

$$P(s) = a_n s^n + a_{n-1}s^{n-1} + ...a_1 s^1 + a_0 \tag{5.14}$$

Based on the cases mentioned, it can be asserted that $P(s)$ will be stable if and only if all the roots of its characteristic polynomial have negative real values.

Consequently, the system Ω will be stable if and only if $P(s)$ is stable, meaning that all its roots reside in the left-half plane of the complex plane ($\mathbb{C}^-$). Furthermore, an essential requirement for the stability of $P(s)$ is that all its coefficients share the same sign.

5.5 Routh–Hurwitz Stability Test

One of the commonly employed techniques for analyzing the stability of linear systems in the time domain is through the use of a tool known as the Routh table. The Routh table is constructed based on the coefficients, denoted as a_i, of a polynomial. The first two rows of the table contain these coefficients, and to create symmetry, any zeros on the right-hand side are replaced. The remaining $(n-1)$ rows of the Routh table can be calculated using the following procedure:

$$
b_1 = \frac{a_{n-1}a_{n-2} - a_n a_{n-3}}{a_{n-1}}, \quad b_2 = \frac{a_{n-1}a_{n-4} - a_n a_{n-5}}{a_{n-1}}, \quad \vdots
$$
$$
c_1 = \frac{b_1 a_{n-3} - a_{n-1}b_2}{b_1}, \quad c_2 = \frac{b_1 a_{n-5} - a_{n-1}b_3}{b_1}, \quad \cdots \tag{5.15}
$$
$$
\vdots \qquad\qquad \vdots \qquad\qquad \ddots
$$

Table 5.2 illustrates the general for of a standard Roth table.

The number of roots in $P(s)$ with positive real parts is equal to the number of sign changes in the first column of the Routh table. For instance, if the first column of the Routh table contains the order (1, 12, −4, 3, 2, and −1), the number of sign changes will be 3 and so, $P(s)$ has three roots with a positive integer. The term a_n is assumed to be normalized to 1 and Routh's vector, by vectoring from all the first-row columns of the Routh table after removing the first input,

$$
r = \left(a_{n-1}, b_1, \ldots, h_1\right)^T \tag{5.16}
$$

Table 5.2 Routh table.

a_n	a_{n-2}	a_{n-4}	$\ldots 0$
a_{n-1}	a_{n-3}	a_{n-5}	$\ldots 0$
b_1	b_2	b_3	$\ldots 0$
c_1	c_2	c_3	$\ldots 0$
$\ldots$	$\ldots$	$\ldots$	$\ldots 0$
g_1	0		
h_1			

$P(s)$ will be stable if and only if $r \succ 0$ and will be unstable if and only if there is one array of r_i that exists, where $r_i \leq 0$. For further details, see [2].

5.6 Kharitonov's Theorem (Interval Routh–Hurwitz Stability Test)

By considering some uncertainties in the system, these uncertainties are applied to the polynomial coefficients. As noted in the previous section, a polynomial with a vector of uncertain coefficients is defined as follows:

$$p(s, a) = a_0 + a_1 s + \ldots + a_n s^n \tag{5.17}$$

$$a := [a_0 + a_1 + \ldots + a_n]^T \tag{5.18}$$

If a is considered in the box defined below, then the polynomial will be called *interval polynomial*:

$$A := \{ a \mid a_i \in [\underline{a}_i, \overline{a}_i], i = 0, 1, \ldots, n \} \tag{5.19}$$

For an interval polynomial, uncertain parameters are defined with polynomial coefficients δ_1, δ_2. An interval polynomial produces the following polynomials:

$$p(s, A) = \{ p(s, a) \mid a \in A \} \tag{5.20}$$

5.6.1 Kharitonov Polynomial Theory

In 1978, Russian researcher Vladimir Kharitonov proved that an interval continuous polynomial would be stable if and only if, by considering the following polynomials:

$$p(s, a) = \{ p(s, a) = a_0 + a_1 s + \ldots + a_n s^n \mid a \in A \}$$
$$a_n > 0 \tag{5.21}$$

In which, $a_i \in [\underline{a}_i, \overline{a}_i], i = 0, 1, 2, \ldots, n$, the following four polynomials are stable:

$$\begin{aligned}
p^{+-}(s) &= \overline{a}_0 + \underline{a}_1 s + \underline{a}_2 s^2 + \overline{a}_3 s^3 + \overline{a}_4 s^4 + \cdots \\
p^{++}(s) &= \overline{a}_0 + \overline{a}_1 s + \underline{a}_2 s^2 + \underline{a}_3 s^3 + \overline{a}_4 s^4 + \cdots \\
p^{-+}(s) &= \underline{a}_0 + \overline{a}_1 s + \overline{a}_2 s^2 + \underline{a}_3 s^3 + \underline{a}_4 s^4 + \cdots \\
p^{--}(s) &= \underline{a}_0 + \underline{a}_1 s + \overline{a}_2 s^2 + \overline{a}_3 s^3 + \underline{a}_4 s^4 + \cdots
\end{aligned} \tag{5.22}$$

In general, with the help of this method, instead of examining the infinitesimal states which increase with increasing the uncertainties, only four polynomials are studied using classical methods, such as the Routh–Hurwitz, the stability of the system is specified.

The above polynomials are called *Kharitonov polynomials* [3]. The overbar and underbar show the upper and the lower intervals of the polynomial coefficients. Various proofs have been made for the Kharitonov method like [4]. Some most well-known examples are *value sets, zero Execution,* and *Mikhailov's Stability Criterion* [5].

Example 5.1   Interval Stability Analysis of an Interval Closed Loop System
Consider the closed-loop system in Figure 5.5. The purpose is to achieve the stable area for the gain K so that we have $a \in [1, 2]$ and $b \in [0.2, 0.8]$.
The closed-loop characteristic equation is $s^3 + [a]s^2 + [b]s + 1 + k = 0$.
To obtain the stability region using the Kharitonov polynomial theory,

$$[a_0] = [1 + k, 1 + k]$$
$$[a_1] = [0.2, 0.8] \tag{5.23}$$
$$[a_2] = [1, 2]$$

By generation of the Kharitonov polynomial,

$$p^{+-}(s) = 1 + k + 0.2s + s^2 + s^3$$
$$p^{++}(s) = 1 + k + 0.8 + s^2 + s^3$$
$$p^{-+}(s) = 1 + k + 0.8s + 2s^2 + s^3 \tag{5.24}$$
$$p^{--}(s) = 1 + k + 0.2s + 2s^2 + s^3$$

Finally, by solving the four equations above using the Routh–Hurwitz method, the control signal is achieved as $k \in [-1, -0.8]$.

Example 5.2   Interval Stability Analysis of a Fourth Order System
To make the concept of stability clearer, assume that the system's interval characteristic parameters as $[a]s^4 + [b]s^3 + [c]s^2 + [d]s + [e] = 0$, from which

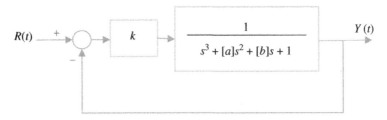

Figure 5.5   The interval closed loop system.

$[a] = [13, 15]$, $[b] = [22, 30]$, $[c] = [28, 32]$, $[d] = [15, 20]$, $[e] = [5, 8]$. The Kharitonov polynomials in this case are

$$p^{+-}(s) = 15s^4 + 22s^3 + 28s^2 + 20s + 8$$

$$p^{++}(s) = 15s^4 + 30s^3 + 28s^2 + 15s + 8$$

$$p^{-+}(s) = 13s^4 + 30s^3 + 32s^2 + 15s + 5 \qquad (5.25)$$

$$p^{--}(s) = 13s^4 + 22s^3 + 32s^2 + 20s + 5$$

Kharitonov rectangular for this system is shown in Figure 5.6.

The source is removed from the Kharitonov rectangle; hence, the interval polynomial $p(s, q)$ is robust and stable. Based on Figure 5.6, the following results are obtained:

1) The order of the system is specified.
2) Since the Kharitonov rectangles circumvent the point $(0, 0)$ in the counter-clockwise direction, the system will remain stable.
3) Because of system stability, optimal control of the system can be used to keep the system stable.

The above method is one of the methods for checking the stability of the interval systems in the time domain. There are also different other methods in the frequency domain (like the interval Nyquist method) that exhibit acceptable results [6–8].

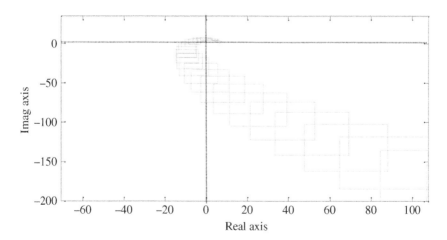

Figure 5.6 The set value of the process model when ω changes from 0 to 7.4 radians per second.

5.6.2 A Centered Representation of the Interval Routh–Hurwitz Stability Criterion

Considering the previously presented interval characteristic polynomial:

$$
\begin{aligned}
p(s) &= [a_0] + [a_1]s + \dots + [a_n]s^n \\
&= (a_{0c} + a_{0r}I_c) + (a_{1c} + a_{1r}I_c)s + \dots + (a_{nc} + a_{nr}I_c)s^n
\end{aligned}
\tag{5.26}
$$

Therefore, we can rewrite it as

$$
p(s) = \underbrace{(a_{0c} + a_{1c}S + \dots + a_{nc}S^n)}_{P_c(s)} + \underbrace{(a_{0r} + a_{1r}s + \dots + a_{nr}s^n)}_{P_r(s)}I_c
\tag{5.27}
$$

By extending the Routh–Hurwitz stability criterion to the $P_c(s)$, we can generalize the theorem of stability based on the interval-centered representation using the radius $P_r(s)$.

5.7 Interval Stability Based on Linear Matrix Inequalities

After the introduction of Kharitonov's theorem, numerous researchers began to explore its application in analyzing dynamic systems [9–11]. In this section, a novel method for conducting stability analysis on interval systems is presented. The essence of this method lies in assessing the positive definiteness of the system matrix, which is elaborated upon in the subsequent discussion.

5.7.1 The Positive Matrix of the Interval Matrix

In general, one can say that an interval matrix of $[A]$ is positive definite, provided that each selected matrix $A \in [A]$ is positive. This case (Positive Determination of the interval Matrices) was studied in [12].

A square matrix (whose symmetry feature is not necessary) is called a positive definite if, $f(A) > 0$, i.e., for each $x^T \neq 0$, $x^T A x > 0$. Accordingly, an interval matrix is called a positive definite, if it provides a positive definite for all $A \in [A]$, so it can be said that $\min\{f(A); A \in [A]\} > 0$.

Theorem 5.1 Symmetric Values and Their Interval Descriptions
Consider an interval matrix $[A] = [A_c - A_r, A_c + A_r]$ that is symmetric [13]. If A_c is a positive definite matrix and $\rho(|A_c^{-1}|A_r) < 1$, then the interval matrix $[A]$ will also be positive definite.

Proof: Because A_c is positive definite, it is invertible and it will supply the condition of the assertion theory for being definiteness of the interval matrix[A] ([A] is definite if, for all $A \in [A]$, it will be nonsingular); hence, based on [11], the interval matrix[A] is positive definite.

Theorem 5.2   Positive Definite Values and Interval Descriptions
The interval matrix $[A] = [A_c - A_r, \ A_c + A_r]$ is positive definite if $\rho(\Delta)$ $< \lambda_{\min}(A_c)$ [14].

Proof: According to the definition [161], $\min\{f(A); A\epsilon[A]\} > \lambda_{\min}(A_c) - \rho(A_r) > 0$, therefore, for each $A \in [A]$, $f(A) > 0$ and therefore, the interval matrix $A \in [A]$ will be positive definite.

5.7.2   Stability Analysis of the Interval Systems

Definition 5.1   Interval Hurwitz Stability
A square matrix A is considered Hurwitz stable if all of its singular values are located on the left side of the imaginary axis ($\lambda < 0$). Similarly, an interval matrix [A] is deemed stable if every matrix $A \in [A]$ is stable. The issue of stability for interval matrices is rooted in control theory and pertains to the behavior of invariant linear systems $\dot{x}(t) = Ax(t)$ under interval disturbances.

Theorem 5.3   Symmetry and Positive Definiteness in Matrices
Let us suppose $[A] = [A_c - A_r, A_c + A_r]$ is an interval matrix. In this condition, the following cases are equivalent [14]:

• The interval matrix [A] is symmetric and A_c is stable.
• $\rho(|\ A_c^{-1}\ |\ A_r) < 1$ is positive definite.

Proof: Suppose that $[A_0]$ is positive definite. Consider a singular value, λ, from the interval matrix $A \in [A]$. According to Bendixon's theory, $Re\ \lambda < \left(\frac{1}{2}(A + A^T)\right)$, where $[A] = \frac{1}{2}(A + A^T)$ is symmetric and belongs to $[A]$; consequently, $[-A] \in [A_0]$.

Therefore, $[-A]$ is a positive definite, and hence all of the singular values of $[A]$ will be negative. As a result, all values of A in the interval will be stable.

As is evident from the above definition, the problem lies in the inequality definition, and therefore, the best approach to analyzing these systems is to employ *Linear Matrix Inequalities (LMIs)*.

5.7.3 Linear Matrix Inequalities

In the following, we will discuss an LMI interval representation of the Routh–Hurwitz stability criterion to simplify the problem. A LMI is a symmetric and affine matrix that can be defined as follows:

$$F(x) = F_0 + \sum_{i=1}^{\infty} x_i F_i > 0 \tag{5.28}$$

where $F_i = F_i^T, F_i \in \mathbb{R}^{n \times n}, x \in \mathbb{R}^m$.

By the following definition, $F(x) = F(x)^T$.

In general, it can be said that any symmetric and affine matrix inequality which depends on its variable can be defined by an LMI.

By considering the maximum singular value, where $\bar{\sigma}(A(x)) < 1$, This equation can be written as an LMI as follows:

$$\bar{\sigma}(A(x)) < 1 \Leftrightarrow A(x)A(x)^T < I \Leftrightarrow I - A(x)I^{-1}A(x)^T \succ 0$$
$$\Leftrightarrow \begin{pmatrix} I & A(x) \\ A(x)^T & I \end{pmatrix} \succ 0 \tag{5.29}$$

Based on Theorem 5.3, two criteria need to be fulfilled for interval stability, which we will redefine separately:

A) The interval matrix $[A]$ is symmetric and A_c is stable.

Consider the Lyapunov function as $V(x) = x^T P x$. Each matrix can be formulated as a summation of symmetric and asymmetric definitions as follows:

$$P = \frac{1}{2}\left(P + P^T\right) + \frac{1}{2}\left(P - P^T\right),$$
$$P_1 = \frac{1}{2}\left(P + P^T\right) = P_1^T, \tag{5.30}$$
$$P_2 = \frac{1}{2}\left(P - P^T\right) = -P_2^T.$$

Therefore,

$$P = \frac{1}{2}\left(P + P^T\right) + \frac{1}{2}\left(P - P^T\right),$$
$$V(x) = x^T P x = x^T P_1 x + x^T P_2 x = \underbrace{\frac{1}{2}x^T\left(P + P^T\right)x}_{V_1(x)} + \underbrace{\frac{1}{2}x^T\left(P - P^T\right)x}_{V_2(x)},$$
$$V_2(x) = \frac{1}{2}x^T P x - \frac{1}{2}x^T P^T x = \frac{1}{2}x^T P^T x - \frac{1}{2}x^T P^T x = 0 \Leftrightarrow V(x) = V_1(x) \tag{5.31}$$

i.e., the matrix is symmetric. Considering the Lyapunov stability, in this case, the system is stable if its derivative is negative, that is,

$$\dot{V}(x) = \dot{x}Px + x^T P\dot{x} = x^T A_c^T Px + x^T PA_c x, \tag{5.32}$$

Here, by considering the LMI definition,

$P \succ 0,$

$$\begin{bmatrix} -[1 \ A_c^T] \begin{bmatrix} 0 & P \\ P & 0 \end{bmatrix} \begin{bmatrix} 1 \\ A_c \end{bmatrix} & 0 \\ 0 & P \end{bmatrix} = \begin{bmatrix} -A_c^T P - PA_c & 0 \\ 0 & P \end{bmatrix} \succ 0. \tag{5.33}$$

The above definition shows that the matrix A is Metzler.

Definition 5.2   Metzler Analysis
Matrix A is Metzler if all of its elements outside the original diameter are nonnegative, i.e., $\forall_{i \neq j}, x_{ij} \geq 0$.

B) $\rho(| A_c^{-1} | A_r) < 1$ is positive definite:

In general, the spectral radius has different definitions, each of which can be used to define LMI. Assuming $A = |A_c| \times A_r$:

$$(D1) \quad \rho(A) = \underset{i}{\max} \ \{A \in C^{n \times n} : |\lambda_i|, \lambda_i \in \sigma_i\} \tag{5.34}$$

$$(D2) \quad \rho(A) = \underset{k \to \infty}{\lim} \|A^k\|^{\frac{1}{k}} \\ k : Type \ of \ Norm \tag{5.35}$$

$$(D3) \quad \rho(A) = \overline{\sigma}(A(x)) \tag{5.36}$$

$$(D4) \quad \rho(A) = Trace(A^T(x)A(x)) \tag{5.37}$$

In this book, (D3) is employed to generate LMI.

Definition 5.3   Perron–Frobenius Theorem
Perron–Frobenius theorem: By assuming $t \in \mathbb{R}$ and $A \in \mathbb{R}_+^{n \times n}$,

$$\exists (tI_n - A)^{-1} \geq 0 \Leftrightarrow \rho(A) \prec t \tag{5.38}$$

Therefore, based on the theory and assuming $t = 1$, $\rho(A) \{<\} 1 \Leftrightarrow (I_n - A)^{-1} \geq 0$. Subsequently, based on [15], it can be said that the system is stable if $(A - I)$ is Hurwitz.

As a result, according to the nonnegative hypothesis of matrix A, the matrix $(A - I)$ is Metzler and it can be stated that matrix $A \in \mathbb{R}_+^{n \times n}$ is a Schur matrix if and only if there is a diagonal positive definite matrix, such that $P \succ 0$, that makes the matrix Hurwitz stable. By defining this theory based on LMI,

$$(A - I)^T P + P(A - I) \prec 0 \tag{5.39}$$

Or in a different way, based on Schur's stability,

$$\begin{aligned} &P \succ 0 \\ &\begin{bmatrix} I & (A - I)^T P + P(A - I) \\ I & 0 \end{bmatrix} \succ 0 \end{aligned} \tag{5.40}$$

If the definition (D4) has been employed for the stability of LMI,

$$\bar{\sigma}(A(x)) \prec 1 \Leftrightarrow A(x)A(x)^{-1} \prec I \Leftrightarrow I - A(x)A(x)^{-1} \succ 0 \tag{5.41}$$

i.e., based on Schur's definition,

$$\begin{aligned} &P \succ 0 \\ &\begin{bmatrix} I & A(x) \\ A^T(x) & I \end{bmatrix} \succ 0 \end{aligned} \tag{5.42}$$

Consequently, for the stability of an interval system, both LMIs should be satisfied. In other words, by combining the two expressions mentioned above, a time-interval system is considered stable if the following conditions are met:

$$\begin{aligned} &P \succ 0, \\ &\begin{bmatrix} F_1(P) & 0 \\ 0 & F_2(P) \end{bmatrix} \succ 0. \end{aligned} \tag{5.43}$$

Where

$$\begin{aligned} F_1(P) &= \begin{bmatrix} -A_c^T P - PA_c & 0 \\ 0 & P \end{bmatrix}, \\ F_2(P) &= \begin{bmatrix} (I - A)^T P + P(I - A) & 0 \\ 0 & P \end{bmatrix}. \end{aligned} \tag{5.44}$$

Or,

$$\begin{aligned} F_1(P) &= \begin{bmatrix} -A_c^T P - PA_c & 0 \\ 0 & P \end{bmatrix}, \\ F_2(P) &= \begin{bmatrix} I & (\mid A_c \mid A_r) \\ (\mid A_c \mid A_r)^T & I \end{bmatrix}. \end{aligned} \tag{5.45}$$

5.8 Controllability and Observability

Controllability is a critical prerequisite in control systems, playing a vital role in various control problems such as stabilizing unstable systems through feedback control or optimal control. Controllability assesses a system's capability to traverse its entire configuration space through specific manipulations. In other words, controllability encompasses the ability to achieve the desired objectives within a dynamic system using control inputs.

Control problems involving interval uncertainties are derived from diverse fields including control theory, game theory, operational research, engineering, and natural sciences [16]. Uncertainty-related issues have been extensively studied by researchers for various systems [17, 18]. On the other hand, controllability in control theory is of particular importance and plays an important role in dynamic control systems [19]. However, investigations specifically addressing problems with interval uncertainties are relatively scarce [3, 20]. The ability to analyze systems with desirable characteristics is closely linked to their controllability and observability properties.

Furthermore, in control theory, observability measures the extent to which the internal states of a system can be inferred from its outputs. Observability is the dual concept of controllability, implying that to fully understand the system's internal behavior, it must also be observable. The notion of observability was introduced by Hungarian-American engineer Rudolf Kalman for linear dynamical systems [21].

5.9 Controllability and Observability Based on Interval Criteria

Consider the system of Eq. (5.5). This interval system is controllable if, for any arbitrary values $x(t_0) = X_0$ and parameters $[A] \in \mathbb{IR}^{n \times n}$, $[B] \in \mathbb{IR}^{n \times m}$, there is a control signal, $u(t)$, that transfers the system from X_0 to X_f.

Therefore, the desired interval system is controllable if for each selected interval state, $A \in [A]$ and control $B \in [B]$, in a considered interval dynamic system, the pair $([A], [B])$ is controllable [22]. We construct a controllable matrix for the pair:

$$\left\{ D \in \mathbb{R}^{n \times mn} \mid [D] = \left[[B] \mid [A][B] \mid \dots \mid [A]^{n-1}[B] \right], \forall A \in [A], \forall B \in [B] \right\}$$

$$(5.46)$$

From the equation above, for instance, the multiplication of two interval matrices $[V] = \left([v]_{ij} \right) \in \mathbb{IR}^{n \times l}$ and $[U] = \left([u]_{ij} \right) \in \mathbb{IR}^{l \times m}$ is an interval matrix $[W] = \left([w]_{ij} \right) \in \mathbb{IR}^{n \times m}$ such that

$$[w]_{ij} = \sum_{k=1}^{l} [v_{ik}][u_{kj}] \tag{5.47}$$

And, the multiplication of the interval parameters, $[u]$, $[v]$, based on interval analysis is

$$[u] \times [v] = [\min\{\underline{uv}, \underline{u}\bar{v}, \bar{u}\underline{v}, \bar{u}\bar{v}\} \tag{5.48}$$

From the conception of controllability, the rank of the controllability matrix should be equal to the rank of the state parameter (n) [85].

By extending this conception to the interval systems and for equalizing the rank of the interval matrix $[D]$ to the $\{n\}$, the tank of each selected $D \in [D]$ should have the same rank, n; i.e.,

$$rank[D] = n \Leftrightarrow \{D \in \mathbb{R}^{n \times mn} \mid rankD = n, \forall D \in [D]\} \tag{5.49}$$

To clarify the general conception of controllability, consider the following example.

Example 5.3   Model of a DC Motor with Interval Uncertainties
DC motors have long been widely used as primary actuators in various industrial applications due to their straightforward characteristics and stability [23]. The speed of a DC motor is directly proportional to the applied voltage. Different techniques, such as electronic controllers and battery trapping, can be employed for controlling the speed of DC motors [24, 25].

However, one crucial problem often overlooked in the mathematical modeling of DC motors is the neglect of uncertain factors, such as variations in resistance values due to temperature changes. This oversight necessitates the design of a controller that does not fully account for system uncertainties. To address this issue, interval arithmetic is utilized to incorporate uncertainties in motor identification. This approach enhances the robustness and practicality of the controller in the face of motor condition changes. The schematic diagram of a DC motor with interval parameters $(\sim)$ is depicted in Figure 5.7.

The motor toque $(\tilde{T}_m)$ is proportional to the armature current $(\tilde{i}_a)$ by the following formula:

$$\tilde{T}_m = \tilde{K}_i \tilde{i}_a, \tag{5.50}$$

The back emf $(\tilde{e}_b)$ is also relative to angular velocity $(\tilde{\omega}_m)$ by:

$$\tilde{e}_b = \tilde{K}_b \omega_m = \tilde{K}_b \frac{d\theta}{dt}, \tag{5.51}$$

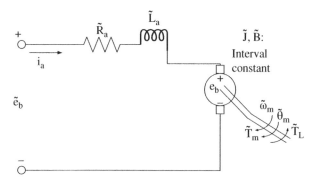

Figure 5.7 Schematic of a DC motor with interval parameters.

By applying Newton's law and Kirchoff's law, the system equations can be achieved as follows:

$$
\tilde{L}_a \frac{\mathrm{d}\tilde{i}_a}{\mathrm{d}t} + \tilde{R}_a \tilde{i}_a = \tilde{e}_a - \tilde{K}_b \frac{\mathrm{d}\theta}{\mathrm{d}t},
$$

$$
\tilde{J}_m \frac{\mathrm{d}^2\theta}{\mathrm{d}t^2} + \tilde{B}_m \frac{\mathrm{d}\theta}{\mathrm{d}t} = \tilde{K}_i \tilde{i}_a.
$$

(5.52)

$$
\tilde{e}_b = \tilde{K}_b \omega_m = \tilde{K}_b \frac{\mathrm{d}\theta}{\mathrm{d}t},
$$

(5.53)

From the equations above, the state space of the system can be considered as follows:

$$
\begin{bmatrix} \dot{\tilde{i}}_a \\ \dot{\tilde{\omega}}_m \\ \dot{\tilde{\theta}}_m \end{bmatrix} = \begin{bmatrix} -\tilde{R}_a/\tilde{L}_a & -\tilde{K}_b/\tilde{L}_a & \{0\} \\ \tilde{K}_i/\tilde{J}_m & -\tilde{B}_m/\tilde{J}_m & \{0\} \\ \{0\} & \{1\} & \{0\} \end{bmatrix} \begin{bmatrix} \tilde{i}_a \\ \tilde{\omega}_m \\ \tilde{\theta}_m \end{bmatrix} + \begin{bmatrix} 1/\tilde{L}_a \\ \{0\} \\ \{0\} \end{bmatrix} \tilde{e}_a,
$$

$$
\tilde{\omega}_m = \begin{bmatrix} \{0\} & \{1\} & \{0\} \end{bmatrix} \begin{bmatrix} \tilde{i}_a \\ \tilde{\omega}_m \\ \tilde{\omega}_m \end{bmatrix}.
$$

(5.54)

where the terms $\{0\} = [0, 0]$ and $\{1\} = [1, 1]$ are degenerate integers [26]. The interval standard test model with interval parameters is considered and listed in Table 5.3.

Consider Eq. (5.28). The system is controllable if, for the initial condition $x(0) = x_0$, and for any given vector x_f, there exists a finite time interval $[0, t_f]$ and an input function $u(t)$ within this interval that can map the system from x_0 to x_f within time t_f, i.e., $x(t_f) = x_f$. Conversely, if such a mapping is not possible, then the system is uncontrollable. Various methods exist for analyzing the

Table 5.3 DC motor parameters with interval values.

Symbol	Value with interval uncertainty
$\tilde{E}$	[11, 13] (volt)
$\tilde{J}_m$	[0.008, 0.02] (kgm^2)
$\tilde{B}_m$	[0.00002, 0.00005] (kgm^2/s)
$\tilde{K}_i$	[0.021, 0.024] (Nm/A)
$\tilde{K}_b$	[0.020, 0.024] (V/rad/s)
$\tilde{R}_a$	[0.5, 1.5] (Ω)
$\tilde{L}_a$	[0.25, 0.75] (H)

controllability of systems with real-valued variables [27]. Consider the introduced third-order interval DC model with the following interval parameters:

$$
\tilde{A} = \begin{bmatrix} [-6, -0.6] & [-0.096, -0.027] & \{0\} \\ [1.05, 3] & [-0.0062, -0.001] & \{0\} \\ \{0\} & \{1\} & [\{0\}] \end{bmatrix}
$$
$$
\tilde{B} = \begin{bmatrix} 4/3, 4 \\ \{0\} \\ \{0\} \end{bmatrix}, \tilde{C} = [\{0\} \ \ \{1\} \ \ \{0\}]
$$

(5.55)

The system $(\tilde{A}, \tilde{B})$ will be controllable if the controllability matrix $\tilde{R} = \begin{bmatrix} \tilde{B} \tilde{A} \tilde{B} \tilde{A}^2 \tilde{B} \tilde{A}^{n-1} \tilde{B} \end{bmatrix}$ has full row rank (i.e., $\rho(\tilde{R}) = \{n\} = [n, n]$). By extending the interval arithmetic to this assumption and calculating the system controllability matrix,

$$
\rho(\tilde{R}) = \rho \left(\begin{bmatrix} [1.33, 4] & [-24, 6.4] & [-86.74, 144.15] \\ \{0\} & [-1.2, 12] & [-72.07, 36.39] \\ \{0\} & \{0\} & [-1.20, 12] \end{bmatrix} \right) = \{3\}
$$

(5.56)

Based on the analysis above, it is evident that the studied system is controllable, allowing for the placement of the closed-loop poles at any desired location in the s-plane. Next, we present a developed method for analyzing the controllability of systems with interval uncertainties.

Definition 5.4   Interval Controllability

The system of Eq. (5.4) (pairs $(C, [A])$) is controllable if and only if

$$rank[D] = n \tag{5.57}$$

According to the explanations, the observability matrix will be also as follows.

$$[\Omega] = \begin{bmatrix} C \\ C[A] \\ \vdots \\ C[A]^{n-1} \end{bmatrix} = \left\{ \Omega \in \mathbb{R}^{ln \times n} \middle| \Omega = \begin{bmatrix} C \\ C[A] \\ \vdots \\ C[A]^{n-1} \end{bmatrix}, \forall A \in [A] \right\} \tag{5.58}$$

In this case, the observability of the interval dynamic system is obtained by the following definition.

Definition 5.5   Interval Visibility

The open system (4.4) (pairs) will be visible if and only if

$$rank[\Omega] = n \tag{5.59}$$

The accuracy of the two expressions described for controllability and observability is achieved when considering all possible values within the defined interval. To simplify the calculations involving the interval matrix, singular value-based methods have been employed [28].

5.9.1   Singular Values for Analyzing the Controllability

The smallest singular value is a defined value for examining the nonsingularity of a normal matrix. A singular decomposition for a matrix $D \in \mathbb{R}^{n \times mn}$ with order k is as follows:

$$D = V \sum W \tag{5.60}$$

where $V \in R^{n \times n}$, $W \in R^{mn \times mn}$ are orthogonal matrices whose columns have eigenvalues DD^T and $D^T D$, respectively.

The matrix contains a diagonal square. The diagonal elements of a matrix are called singularity of the matrix, which are nonnegative square roots of special matrix values.

The matrix $\sum = (\sigma_{sj}) \in R^{n \times mn}$ includes $n \times n$ diagonal square, in which $\sigma_{sj} = 0$ for $s \neq j$ and $\sigma_{11} \geq \sigma_{22} \geq ...\sigma_{ss}... \geq \sigma_{k+1, k+1} = ...\sigma_{nn} = 0$ and $n \neq mn$ for extra zero columns.

The diagonal elements $\sigma_s \equiv \sigma_{ss}$ of the matrix Σ, also referred to as the singular values of the matrix D, represent the nonnegative square roots of the singular values of the matrix $[a_{ij}] = [\underline{a}_{ij}, \overline{a}_{ij}]$. The matrix D is considered full rank if and only if its minimum singular value is positive, denoted as $\sigma_{\min}(D) > 0$. One approach for determining the rank of a matrix involves calculating its singular decomposition and counting the number of singular values. With the aforementioned definition in mind, the controllability matrix can be obtained using the following method outlined below.

Definition 5.6   Interval Singularity
The following set that includes the singular values of the total real-valued matrices $D \in [D]$ will be called the singular values of the interval matrix $[D] \in \mathbb{IR}^{n \times mn}$.

$$[\sigma]_s([D]) = [\underline{\sigma}_s([D]); \overline{\sigma}_s([D])] \Leftrightarrow \{\sigma_s(D), s = 1, ..., n | \forall D \in [D]\}, \tag{5.61}$$

Definition 5.7   Interval Equality of Rank in Matrix
The rank of the matrix $[D] \in \mathbb{IR}^{n \times mn}$ will be equal to k, if and only if the singular values of it to an interval positive set are equal to k:

$$rank \ [D] = k \leftrightarrow \{\sigma_s([D]) \in Z \ \forall s = 1, 2, ..., k, \ k \leq n\} \tag{5.62}$$

where $Z = \{[z] \in \mathbb{IR} | \underline{z} > 0 \,\&\, \overline{z} > 0\}$ is a subset of the positive intervals.

Since the singular values of an interval matrix $[D] \in \mathbb{IR}^{n \times mn}$ are nonnegative square roots of the singular values of the matrix $DD^T \in \mathbb{IR}^{n \times mn}$, the following relation can be used:

$$[\sigma]_s([D]) = \sqrt{\lambda\left([D][D]^T\right)}, s = 1, 2, ..., n. \tag{5.63}$$

Definition 5.8   Positive Definiteness of an Interval Matrix
An interval matrix $[G] \in \mathbb{IR}^{n \times n}$, where its components $[g]_{ij} = [\underline{g}_{ij}, \overline{g}_{ij}]$ are called positive definite, and it is defined as $[G] \succ 0$ if any $G \in [G]$ is positive, i.e., for all $G \in [G]$, the square form $x^T G x \succ 0, \forall x \neq \in \mathbb{R}^n$.

Definition 5.9   Symmetric Definition of an Interval Matrix
An interval matrix is called symmetric if and only if

$$[G] = [G]^T \Leftrightarrow \{G \in \mathbb{R}^{n \times n} \mid \forall G \in [G] = [\underline{G}, \overline{G}], G = \tilde{G}, \underline{G} < G < \overline{G}\}. \tag{5.64}$$

From the equation above, an interval matrix $[D][D]^T$ is positive definite symmetric if $DD^T \in [D][D]^T$ is a positive definite symmetric.

Based on the problem-solving of the eigenvalues of the symmetric interval matrices [35], the eigenvalues of the interval matrix $[D][D]^T$ are achieved as follows:

$$\lambda_s\left([D][D]^T\right) = \left[\lambda_s\left\{med\left([D][D]^T\right)\right\} - \varepsilon_s; \lambda_s\left\{med\left([D][D]^T\right)\right\} + \varepsilon_s\right],$$
(5.65)

where $med\left([D][D]^T\right) = \left(\overline{DD}^T + \underline{DD}^T\right)/2$ is the mean value of the matrix, $\lambda_s\{med([D][D]^T)\}$ is the eigenvalue of the matrix $med([D][D]^T)$, and

$$\varepsilon_s = \left\|[D][D]^T v_s - \lambda_s\left\{med\left([D][D]^T\right)v_s\right\}\right\|_2$$
(5.66)

is nullity vector. Here, v_s is an eigenvector for eigenvalue of $\lambda_s\{med([D][D]^T)\}$. The norm of these two vectors is

$$\|[x]\|_2 = \left(\sum_{i=1}^{n} |x|^2\right)^{1/2}, |[x]| = col\left(|[x]_i|\right)_{i=1}^{n}, |[x]_i| = \max\{|\underline{x}_i|, |\overline{x}_i|\}$$
(5.67)

Assume that $\langle[x]\rangle$ is the minimum distance of the interval points $[x]$ from zero [29]:

$$\langle[x]\rangle = \begin{cases} \min\{|\underline{x}_i|, |\overline{x}_i|\}, & \text{if } 0 \notin [x] \\ 0, & \text{if } 0 \in [x] \end{cases}$$
(5.68)

Consequently, the controllability analysis steps can be generally considered as follows:

1) Compute controllability matrix $[D]$ for the interval pair($[A]$, $[B]$):

$$[D] = \left[[B] \mid [A][B], ..., [A^{n-1}][B]\right]$$

2) Generate Matrix $[D][D]^T$.
3) Determine the median of $[D][D]^T$.
4) Determine the singular values and vectors of the interval matrix $[D][D]^T$ ($\lambda_i(med([D][D]^T))$, $i = 1, ..., n-1$).
5) Find the null space of the interval matrix $null(med([D][D]^T)) = \varepsilon_i, i = 1, ..., n-1$
6) Find interval singular values:

$$\lambda_i\left([D][D]^T\right) = \left[\lambda_i\left(med\left([D][D]^T\right)\right) - \varepsilon_i, \lambda_i\left(med\left([D][D]^T\right)\right) + \varepsilon_i\right], i = 1, ..., n-1$$

7) If all the interval singular values belong to positive and definite interval sets, the order of the interval set is perfect and the system is controllable.

Example 5.4 Interval Controllability Analysis of a Second-Order System
Consider the following system:

$$[A] = \begin{bmatrix} [2,4] & [1,1] \\ [1,1] & [2,4] \end{bmatrix}, [B] = \begin{bmatrix} [1,3] & 0 \\ 0 & [1,3] \end{bmatrix}. \tag{5.69}$$

The purpose is to analyze the controllability of the system. The first step is to generate the controllability matrix for the pair $([A], [B])$:

$$[D] = [[B] \mid [A][B]] = \begin{bmatrix} [1,3] & 0 & [2,12] & [2,6] \\ 0 & [1,3] & [1,3] & [2,12] \end{bmatrix}. \tag{5.70}$$

To find the singularities of the controllable matrix, the following matrix should be found:

$$[D][D]^T = \begin{bmatrix} [1,3] & 0 & [2,12] & [2,6] \\ 0 & [1,3] & [1,3] & [2,12] \end{bmatrix} \begin{bmatrix} [1,3] & 0 & [2,12] & [2,6] \\ 0 & [1,3] & [1,3] & [2,12] \end{bmatrix}^T$$

$$= \begin{bmatrix} [9,189] & [6,108] \\ 6,108 & [6,162] \end{bmatrix}$$

$$\tag{5.71}$$

Afterward, the median matrix $med([D][D^T])$ can be obtained by

$$med([D][D^T]) = \begin{bmatrix} 99 & 57 \\ 57 & 86 \end{bmatrix} \tag{5.72}$$

The eigenvalues achieved as $\lambda_1(med([D][D^T])) = 35.13$ and $\lambda_2(med([D][D^T])) = 149.87$.

Eigenvectors of the above eigenvalues are $v_1 = \begin{bmatrix} 0.665 \\ -0.745 \end{bmatrix}$ and $v_1 = \begin{bmatrix} 0.665 \\ -0.745 \end{bmatrix}$, respectively.

Therefore, from the equations explained before, the nullity vector for the interval matrix $[D][D^T]$ is

$$\varepsilon_1 = \left\| \begin{bmatrix} [9,189] & [6,108] \\ [6,108] & [6,12] \end{bmatrix} \begin{bmatrix} 0.665 \\ -0.745 \end{bmatrix} - 35.13 \begin{bmatrix} 0.665 \\ -0.745 \end{bmatrix} \right\|_2 = 33.75 \tag{5.73}$$

$$\varepsilon_2 = \left\| \begin{bmatrix} [9,189] & [6,108] \\ 6,108 & [6,12] \end{bmatrix} \begin{bmatrix} 0.745 \\ 0.665 \end{bmatrix} - 149.87 \begin{bmatrix} 0.745 \\ 0.665 \end{bmatrix} \right\|_2 = 136.06 \tag{5.74}$$

Based on the explanation given, the values obtained for the interval matrix $[D]$ $[D^T]$ will be equal to

$$\lambda_1([D][D^T]) = [\lambda_1(med([D][D^T])) - \varepsilon_1, \lambda_1(med([D][D^T])) + \varepsilon_1]$$
$$= [\underline{\lambda_1}, \overline{\lambda_1}] = [1.40, 68.86],$$
(5.75)

$$\lambda_2([D][D^T]) = [\lambda_2(med([D][D^T])) - \varepsilon_2, \lambda_2(med([D][D^T])) + \varepsilon_2]$$
$$= [\underline{\lambda_2}, \overline{\lambda_2}] = [13.81, 285.93].$$
(5.76)

Both singular values for the controllability matrix belong to the positive range. Therefore, the studied system is controllable.

Example 5.5 Interval Controllability Analysis of a Two-Wheeled and Self-Balancing Robot Motor
Consider a two-wheeled and self-balancing robot motor speed model with interval uncertainties as follows [23, 28]:

$$[A] = \begin{bmatrix} \{0\} & \{1\} & \{0\} & \{0\} \\ \{0\} & [-0.25, -0.11] & [24.6, 56.05] & \{0\} \\ \{0\} & \{0\} & \{0\} & \{1\} \\ \{0\} & [-0.61, -0.49] & [237, 239] & \{0\} \end{bmatrix}$$

$$[B] = \begin{bmatrix} \{0\} \\ [0.41, 0.56] \\ \{0\} \\ 1.9, 2.72 \end{bmatrix}, [C] = [\{0\} \quad \{0\} \quad \{1\} \quad \{0\}]$$
(5.77)

The system will be controllable if for the initial condition of $x(0) = x_0$ and for any given vector x_f, there will be a limited time, t_f and input $u(t)$ in the interval $[0, t_f]$, where this input mapping the system from the x_0 into x_f in time t_f, i.e., $x(t_f) = x_f$, in otherwise, the given equation is uncontrollable [27, 28]. Figure 5.8 shows a two-wheeled and self-balancing robot.

From the explanations before, for the case study, we have

$$[D] = [[B] \mid [A][B] \mid [A]^2[B] \mid [A]^3[B]]$$

$$= \begin{bmatrix} \{0\} & [0.41, 0.56] & [-0.14, -0.045] & [46.74, 152.49] \\ [0.41, 0.56] & [-0.14, -0.0451] & [46.74, 152.46] & [-57.27, -10.09] \\ \{0\} & [1.9, 2.72] & [-0.34, -0.20] & [450.32, 650.16] \\ 1.9, 2.72 & [-0.34, -0.2] & [450.32, 650.16] & [-174.66, -70.51] \end{bmatrix}.$$
(5.78)

Figure 5.8 Two-wheeled and self-balancing robot.

$$[D][D^T] = \begin{bmatrix} [2185,23254] & [-8754.5,-4737] & [21049,91440] & [-26725,-3316] \\ [-8754.5,-473.7] & [2287,26529] & [-37287,-4553] & [21760,109150] \\ [21049,99144] & [-37287,-4553] & [202790,422720] & [-113780,-31840] \\ [-26640,-3316] & [21760,108830] & [-113780,-31840] & [207760,453220] \end{bmatrix}.$$

$$(5.79)$$

By achieving the mean value,

$$med([D][D^T]) = \begin{bmatrix} 12720 & -6746 & 56245 & -15021 \\ -4614 & 14408 & -20920 & 65455 \\ 60097 & -20920 & 312755 & -72810 \\ -14978 & 65295 & -72810 & 330490 \end{bmatrix}.$$

$$(5.80)$$

After applying the formula in the flowchart, the eigenvalues and the null values are achieved as follows:

$$\lambda_2 = 255930; \; \varepsilon_2 = 2.8086e^{-10} \quad \lambda_1 = 411360; \; \varepsilon_1 = 3.6007e^{-10}$$
$$\lambda_4 = 470; \; \varepsilon_4 = 4.7749e^{-11} \quad \lambda_3 = 3170; \; \varepsilon_3 = 1.6130e^{-11}$$

And the final interval eigenvalues are $\tilde{\lambda}_1 = \{411360\}$, $\tilde{\lambda}_2 = \{3170\}$, $\tilde{\lambda}_3 = \{255930\}$, and $\tilde{\lambda}_4 = \{470\}$. As can be seen, all of the interval eigenvalues are positive so the system is controllable.

Note that due to the small value of ε and the large values of the eigenvalues, the lower and upper bounds of the interval eigenvalues closely resemble the dominant integers.

5.10 Conclusions

In this chapter, we provided a brief overview of various control methods, including classic control, conventional advanced control, robust control, and adaptive control. The focus now shifts to analyzing the impact of uncertainties on system stability and controllability. It is essential to investigate the stability and controllability of interval systems as a prerequisite for achieving optimal control of interfaces and determining confidence intervals. Interval-based methods are introduced to examine these characteristics in interval devices. Subsequently, a study on sustainability is presented using LMIs based on the Routh–Hurwitz stability criterion. This approach provides a more rigorous analysis of system stability. Finally, interval control, which serves as a crucial prerequisite for optimal control, is examined and applied to various examples.

References

1 Lemos, J.M., *Optimal Linear Quadratic Control. Control Systems, Robotics and Automation–Volume VIII: Advanced Control Systems-II*. 2009: Encyclopedia of Life Support Systems (EOLSS). p. 124.

2 Corriou, J.-P., *Stability analysis*, in *Process Control*. 2018, Springer. p. 117–142.

3 Yang, X., et al., *Robust stability analysis of active voltage control for high-power IGBT switching by Kharitonov's theorem*. IEEE Transactions on Power Electronics, 2016. **31**(3): p. 2584–2595.

4 Minnichelli, R.J., J.J. Anagnost, and C.A. Desoer, *An elementary proof of Kharitonov's stability theorem with extensions.* IEEE Transactions on Automatic Control, 1989. **34**(9): p. 995–998.

5 Mori, T. and H. Kokame. *Extended Kharitonov's theorems and their application*, in *American Control Conference*, 21–23 June 1989. 1989, Pittsburgh, PA, USA, IEEE.

6 Lee, C.-H., *Sufficient conditions for robust stability of discrete large-scale interval systems with multiple time delays.* Journal of Applied Mathematics and Physics, 2017. **1**(2): p. 1–2.

7 Yu, Y. and Z. Wang, *A graphical test for the interval stability of fractional-delay systems.* Computers & Mathematics with Applications, 2011. **62**(3): p. 1501–1509.

8 Zhang, W. and L. Xie, *Interval stability and stabilization of linear stochastic systems.* IEEE Transactions on Automatic Control, 2009. **54**(4): p. 810–815.

9 Swain, S. and P. Khuntia, *Kharitonov based robust stability for a flight controller.* International Journal of Systems Signal Control and Engineering Applications, 2014. **7**(2): p. 26–32.

10 Pastravanu, O. and M. Voicu, *Necessary and sufficient conditions for componentwise stability of interval matrix systems.* IEEE Transactions on Automatic Control, 2004. **49**(6): p. 1016–1021.

11 Lin, C., et al., *Analysis on robust stability for interval descriptor systems.* Systems & Control Letters, 2001. **42**(4): p. 267–278.

12 Skalna, I., *Positive definiteness and stability of parametric interval matrices.* arXiv preprint arXiv:1709.00853, 2017.

13 Neumaier, A., *Interval Methods for Systems of Equations.* Vol. 37. 1990: Cambridge University Press.

14 Rohn, J., *Positive definiteness and stability of interval matrices.* SIAM Journal on Matrix Analysis and Applications, 1994. **15**(1): p. 175–184.

15 Twardy, M., *An LMI approach to checking stability of 2D positive systems.* Technical Sciences, 2007. **55**(4). p. 385.

16 Ashchepkov, L., *The controllability of an interval linear discrete system.* Journal of Computer and Systems Sciences International, 2007. **46**(3): p. 399–406.

17 Ahn, H.-S., K.L. Moore, and Y. Chen. *Linear independency of interval vectors and its applications to robust controllability tests.* in *Decision and Control, 2005 and 2005 European Control Conference. CDC-ECC'05. 44th IEEE Conference on.* 2005. IEEE.

18 Ashchepkov, L.T., *External bounds and step controllability of the linear interval system.* Automation and Remote Control, 2008. **69**(4): p. 590–596.

19 Rosenbrock, H.H., *State-space and multivariable theory.* IEEE Transactions on Systems, Man, and Cybernetics, 1972. **SMC-2**(2).

20 Zhirabok, A., *Analysis of controllability degree of discrete dynamic system.* Journal of Computer and Systems Sciences International, 2007. **46**(2): p. 169–176.

21 Kalman, R., *On the general theory of control systems.* IRE Transactions on Automatic Control, 1959. **4**(3): p. 110–110.

22 Deng, J., et al., *A novel power market clearing model based on the equilibrium principle in microeconomics.* Journal of Cleaner Production, 2017. **142**: p. 1021–1027.

23 Razmjooy, N. and M. Ramezani, *Optimal control of two-wheeled self-balancing robot with interval uncertainties using Chebyshev inclusion method.* Majlesi Journal of Electrical Engineering, 2018. **12**(1): p. 13–21.

24 Huang, J.T., *Persistent excitation in a shunt DC motor under adaptive control.* Asian Journal of Control, 2007. **9**(1): p. 37–44.

25 Chu, H., et al., *Low-Speed Control for Permanent-Magnet DC Torque Motor Using Observer-Based Nonlinear Triple-Step Controller.* IEEE Transactions on Industrial Electronics, 2017. **64**(4): p. 3286–3296.

26 Sola, H.B., et al., *Interval type-2 fuzzy sets are generalization of interval-valued fuzzy sets: toward a wider view on their relationship.* IEEE Transactions on Fuzzy Systems, 2015. **23**(5): p. 1876–1882.

27 Pappas, G.J., G. Lafferriere, and S. Sastry, *Hierarchically consistent control systems.* IEEE Transactions on Automatic Control, 2000. **45**(6): p. 1144–1160.

28 Shashikhin, V., *Robust stabilization of linear interval systems.* Journal of Applied Mathematics and Mechanics, 2002. **66**(3): p. 393–400.

29 Shary, S.P., *Algebraic approach in the "outer problem" for interval linear systems.* Fundamentalnaya i Prikladnaya Matematika, 2002. **8**(2): p. 567–610.

6

Optimal Control of the Systems with Interval Uncertainties

6.1 Introduction

There is a vast amount of literature available on the topic of defining optimal control problems (OCPs). Numerous books have been published that provide detailed coverage of this subject, including Kirk's [1] and Naidu's [2]. These books are widely recognized as important resources for individuals seeking a deeper understanding of OCPs. One of the key aspects emphasized in these books is the utilization of the Euler–Lagrange and Pontryagin methods. These methods are extensively employed in the field of optimal control and play a crucial role in solving complex optimization problems.

The Euler–Lagrange method involves determining the necessary conditions that must be satisfied by the optimal solution to a given problem. On the other hand, the Pontryagin method is utilized to ascertain the optimal control and state trajectories for a particular system. These methods provide a rigorous framework for analyzing and solving OCPs and are indispensable components of any comprehensive study in this area.

In the background of optimal control with a limited time horizon, the general problem is to select an acceptable control signal, denoted as u^*, to guide the system, described by $\dot{x}(t) = g(x(t), u(t), t)$, along an acceptable path, represented by x^*. Simultaneously, the objective is to minimize the following cost function:

$$J(x, u) = S(x(t_f), t_f) + \int_{t_0}^{t_f} f(x(t), u(t), t)dt \tag{6.1}$$

In the above equation, u^* is the optimum control signal and x^* is the optimal path curve. In the case of binding and having boundary conditions (constraints), the following considerations should also be considered:

$$\varphi(x(t_0), t_0, x(t_f), t_f) = 0 \tag{6.2}$$

Interval Analysis: Application in the Optimal Control Problems, First Edition. Navid Razmjooy.
© 2024 The Institute of Electrical and Electronics Engineers, Inc.
Published 2024 by John Wiley & Sons, Inc.

The path vector, $x(.)$ and the control vector $u(.)$ are in the space $\mathbb{R}^n$ and $\mathbb{R}^m$, respectively; therefore, the path and the control vectors can be considered as follows:

$$x(t) \triangleq \begin{bmatrix} x_1(t) \\ \vdots \\ x_2(t) \end{bmatrix}, u(t) \triangleq \begin{bmatrix} u_1(t) \\ \vdots \\ u_2(t) \end{bmatrix}. \tag{6.3}$$

The functional mapping $S(x(t_f), t_f)$, $f(x(t), u(t), t)$, and $\varphi(x(t_0), t_0, x(t_f), t_f)$ are assumed as follows:

$$S\big(x(t_f), t_f\big) : \mathbb{R}^n \times \mathbb{R} \to \mathbb{R}$$

$$f(x(t), u(t), t) : \mathbb{R}^n \times \mathbb{R}^m \times \mathbb{R} \to \mathbb{R}$$

$$\varphi\big(x(t_0), t_0, x(t_f), t_f\big) : \mathbb{R}^n \times \mathbb{R} \times \mathbb{R}^n \times \mathbb{R} \to \mathbb{R}^q$$

When it comes to solving OCPs, two main strategies can be employed: direct methods and indirect methods. Direct methods involve discretizing the control and state variables and then solving the resulting nonlinear programming problem using numerical optimization techniques. On the other hand, indirect methods involve deriving necessary conditions for optimality using the calculus of variations or Pontryagin's maximum principle. The resulting set of differential equations, referred to as the "adjoint equations," are then solved backward in time [3]. Several specific methods fall under the categories of direct and indirect methods:

1) Belman's Dynamic Programming method, which is based on the Hamiltonian–Jacobi–Belman (HJB) equation.
2) The method of calculus of variations and Pontryagin's maximum principle involves the use of the Euler–Lagrange equations.
3) Direct methods using parameterization and discretization, which utilize non-linear mathematical programming techniques.

For example, in the case of Belman's dynamic programming method, the HJB problem is solved based on closed-loop control. Other methods, such as the calculus of variations and Pontryagin's maximum principle, convert the OCP into a two-point boundary value problem (TPBVP). The resulting optimal control from these methods is ultimately implemented as a closed-loop control strategy. These views are primarily considered indirect methods.

On the other hand, methods that are based on parameterization or discretization of the OCP are known as direct methods. These methods operate as open-loop controls and solve the OCP by converting it into a nonlinear programming problem. In this book, both direct and indirect methods will be used desirably for uncertain systems.

6.2 Indirect Methods

An indirect method involves transforming the problem before solving it. As mentioned earlier, these methods rely on solving HJB equations or finding solutions that convert OCPs into TPBVP problems. In indirect methods, a double-valued boundary value problem (BVP) is considered based on the principle of minimization and other optimal conditions.

Due to the inability to solve these equations analytically, numerical methods like the shooting method are used. Additionally, spectral methods and orthogonal methods can be applied to address the aforementioned problem, as referenced in [4]. There are several specific indirect methods:

1) *Power series approach*: This method seeks an approximate solution for the HJB or TPBVP equation using power series expansions. The resulting approximate feedback law has shown promising results.
2) *Extended linearization method*: In this approach, the nonlinear dynamical system is expressed as a quasilinear form $\dot{x} = f(x(t), u(t), t)$, where $f(x(t), u(t), t)$ is a nonlinear function in x.
3) *Inverse OCP*: This method involves closed-loop optimal control, with the primary objective being the calculation of a cost function based on the optimal trajectory and control selection [5].

Indirect methods offer the advantage of providing accurate measurements and ensuring optimality. However, they also have some disadvantages:

- These methods have a limited convergence domain. If the initial guess is not suitable, convergence may not be achieved.
- Designers need to possess a good understanding of the physical nature of the problem in order to provide an appropriate initial guess, which is not always feasible.
- Extracting the optimal conditions in indirect methods can be challenging.

6.3 Direct Methods

The direct method is a prominent approach for solving OCPs. Unlike the indirect method, which involves transforming the problem before solving it, the direct method incorporates the placement of path and control parameters directly into the cost function to determine the optimal solution. This approach effectively converts the dynamic OCP into a static parametric optimization problem.

One key advantage of the direct method is that it simplifies the problem by avoiding the explicit construction of the Hamiltonian system. Additionally, there are various techniques and software available for solving OCPs using direct methods, making them a popular choice in practice.

There are two main approaches for implementing direct methods: discretization and parameterization.

Discretization entails dividing the time interval into a finite number of subintervals. The optimal solution at each subinterval is then approximated using a set of basis functions. This approach allows for a more accurate representation of the optimal solution, but it can lead to a larger number of decision variables and computational complexity.

Parameterization involves approximating the control and state variables as functions of a set of parameters. These parameters are then optimized directly to find the optimal solution. Parameterization can simplify the problem by reducing the number of decision variables, but it may introduce some level of approximation error.

The choice between discretization and parameterization depends on the specific problem and the researcher's preferences. Both methods have their strengths and weaknesses.

Overall, the direct method is a powerful tool for solving OCPs. Its popularity stems from its ability to transform complex dynamic problems into simpler static ones. By utilizing discretization or parameterization, direct methods offer an efficient and effective way to solve a wide range of OCPs.

- Discretization

Discretization is the process by which the interval $t \in [t_0, t_f]$ is divided into n equal periods, i.e.,

$$t_0 < t_1 < ... < t_n = t_f \tag{6.4}$$

Consequently, based on the discretization method, the variables are sampled at each time point in the above formula.

In general, there are two methods for discretization in optimal control: discretization based on path and control and discretization based on control [6].

- Parameterization

Parameterization is a technique commonly employed in optimal control to approximate functions or variables using a set of known functions, which include adjustable parameters. This approach proves particularly beneficial when directly solving the OCP becomes too complex or when the dimensionality of the problem is high. There are several methods for parameterization, including control parametrization, path parameterization, and a combination of both.

In control parametrization, the control input is approximated using a set of basis functions that possess adjustable parameters. The resulting optimization problem is then solved using numerical techniques, such as nonlinear programming. This approach is especially useful when the control input serves as the primary unknown, with the state variables being considered a function of the control input.

Conversely, path parameterization involves approximating the path or trajectory of the state variables using a set of basis functions with adjustable parameters. This technique proves beneficial when the state variables are the primary unknowns, and the control input is regarded as a function of the state variables.

In more complex cases, a combination of control and path parameterization techniques can be employed to solve OCPs. This approach entails approximating both the control input and state variables using a set of basis functions with adjustable parameters. The resulting optimization problem is then solved using numerical techniques.

By utilizing parameterization techniques, OCPs can be effectively addressed, even when direct solutions are impracticable due to complexity or high dimensionality. These methods allow for efficient approximation and facilitate the use of numerical techniques to find optimal solutions.

1) Control parameterization

In this method, control variables are approximated using a series of finite-length functions with unknown parameters. This definition can be mathematically expressed by the following formula:

$$u_k = \sum_{i=0}^{N} a_i \varphi_i(t) \tag{6.5}$$

where N represents the approximation order, a_i denotes the unknown parameters of the control, and $\varphi_i(t)$ represents properly defined functions that serve as the basis functions of the control space. By integrating the path equation in this case, the state variables are functionally defined as uncertain parameters of the control variables. Both the control and state variables are directly embedded within the equation. Consequently, the complex OCP is transformed into a static optimization problem involving unknown parameters. This method, known as direct methods, is commonly employed; however, it is worth noting that integrating the state equation to obtain the path variables can be computationally expensive.

2) State parameterization

In this method, only the state variable is approximated using a series of finite-length functions with unknown parameters. This state can be mathematically described by the following formula:

$$x_k = \sum_{i=0}^{N} b_i \varphi_i(t) \tag{6.6}$$

Here, b_i represents the unknown parameters of the state variables. The control variable can be derived from the state equations. The key idea behind this method is to select a set of state variables that can be directly approximated by a series of finite-length known functions with unknown parameters. The remaining variables, including the control and other state variables, can then be obtained through the state equations. This approach significantly reduces the overall system size.

3) Control and state parameterization

In this approach, both the control and state variables are approximated using a finite series of known functions with unknown parameters. Mathematically, this state can be defined by the following formula:

$$x_k = \sum_{i=0}^{N} b_i \varphi_i(t) \tag{6.7}$$

$$u_k = \sum_{i=0}^{N} a_i \varphi_i(t) \tag{6.8}$$

Here, x_k represents the state variable, b_i denotes the unknown parameters, and $\varphi_{i(t)}$ represents the chosen basis functions used for approximation.

Using this method, the OCP becomes a nonlinear mathematical optimization problem. Consequently, both the control and the state variables turned into parameters; therefore, the final system will contain a large number of unknown parameters.

• *Advantages of the direct method to indirect method*

In the previous section, we discussed the advantages of using indirect methods, such as the existence and uniqueness of solutions, obtaining exact solutions when the TPBVP has an analytical response, and acquiring measurable error values when numerical values are obtained. However, indirect methods also suffer from several disadvantages, which can be addressed by utilizing direct methods.

One of the drawbacks of indirect methods is that each solution is tailored to a specific problem, necessitating a series of distinct mathematical transformations for each OCP. On the other hand, direct methods offer a more general solution that can be applied to solve any OCP, regardless of its complexity.

Second, in indirect methods, we need to convert the OCP into a single cost function. This means that if we have multiple objectives, we must reduce them into a single function. In contrast, direct methods allow us to employ multiobjective optimization techniques, enabling us to handle problems with multiple conflicting objectives simultaneously.

6.4 Optimal Control Problem in the Presence of Interval Uncertainties

OCPs are often treated as definite problems, where the coefficients and parameters of the model are considered to be known with certainty [7]. However, in reality, there are always uncertainties associated with model parameters due to various factors such as unknown characteristics, unaccounted elements, and time variations. Therefore, it is important to consider these uncertainties in the performance index.

Traditional methods that do not account for uncertain parameters may yield incorrect solutions that are not suitable for proper system control. To address this issue, researchers have explored and developed approaches that consider uncertainties [8]. In recent years, different methods such as the stochastic method [9], fuzzy method [10], and interval analysis [11] have been proposed to handle OCPs with uncertain parameters.

Based on the explanations provided in the first chapter, fuzzy methods can be employed when the probability distributions and membership functions of the system are known. Conversely, random methods are suitable when the system's probability distributions are unknown but the membership functions can be determined. In cases where there is insufficient information about the distribution and membership functions, the interval analysis method is considered the most appropriate [12] Figure 6.1 shows the general configuration of optimal control with interval uncertainties.

Interval analysis can be used for solving OCPs with uncertain parameters only by knowing about the lower and the higher boundaries of the process. By considering these assumptions, the finite horizon OCP can be defined as follows:

$$\min_{u(t), x(t) \,\in\, \Delta} J(X(t), U(t), \Delta) = \int_{t_0}^{t_f} F(t, (X(t), U(t), \Delta)dt, \qquad (6.9)$$

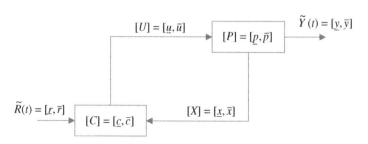

Figure 6.1 General configuration of optimal control with interval uncertainties.

$$\text{subject to}: \dot{X}(t) = G\big(t, (X(t), U(t), \Delta), t \in \big(t_0, t_f\big)$$

$$X(t_0) = X_0$$

$$X\big(t_f\big) = X_f$$

where $X(t)$ and $U(t)$ describe the parameters of the state and the control variables, respectively, $\Delta = [\delta_1, \delta_2, ..., \delta_n]$ is the uncertainties in the system dynamic, F is the interval continuous analytic function of the system, G describes the interval system dynamics, and X_0 and X_f are the initial and final state variables.

Here, $J\big(X(t), U(t), \Delta = \big[\underline{j}(X(t), U(t)), \bar{j}(X(t), U(t))\big]$ is the obtained interval objective function. If the boundary values have constraints,

$$\underline{u} = U = \bar{u}, \underline{x} \leq X \leq \bar{x}, \tag{6.10}$$

$$X(t_0) = X_0, X\big(t_f\big) = X_f$$

Constraints play a crucial role in defining the concept of interval analysis. In this background, interval values represent uncertainties that can introduce variations in the overall solution to an optimization control problem. When dealing with complex problems, the optimal value selected may differ from the original solution, leading to a potential destabilization of the system.

In such cases, it may be more beneficial to prioritize a confident interval for stability rather than selecting an optimal value that may not be correct. By considering a wider range of values within the interval, we can ensure more robust and stable system performance.

6.5 Interval Optimal Control Based on the Indirect Method

In this section, we propose an extension of the indirect optimal control method to handle systems with uncertain parameters. To achieve this, we incorporate interval analysis into the solution process by applying it to the Euler–Lagrange equations.

The Euler–Lagrange equations are fundamental equations in optimal control theory that help determine the optimal control inputs for a given system. By introducing interval analysis to these equations, we can account for uncertainties in the system parameters and obtain an interval-based solution.

Interval analysis allows us to represent uncertain parameters as intervals rather than single values. This approach provides a more robust representation of the uncertainties and enables us to obtain a range of possible solutions instead of a single deterministic solution. By considering the entire interval of possible solutions, we can ensure that the resulting control strategy is robust and capable of handling variations in the system parameters.

By utilizing interval analysis on the Euler–Lagrange equations, we enhance the capability of the indirect optimal control method to handle uncertain parameters. This extension enables us to account for uncertainties and obtain interval-based solutions that are more suitable for real-world systems.

6.5.1 Analysis of the Standard Interval Calculus of Variations

This section represents the extremum response of the functional in interval calculus of variations based on the standard *calculus of variations* lemma. Assume the function $X(t)$ analytic and continuous, i.e., $X(t) \in \mathbb{C}^1$. At first, the following functional optimization is considered as the index performance (PI).

$$\begin{cases} \dot{X}(t) = G(t, X(t), U(t), \Delta) \\ \quad X(t_0) = X_0 \end{cases} \tag{6.11}$$

In the equation above, t_f and $X(t_f)$ can take any value.

In this case, the OCP with interval uncertainties can be defined as follows:

$$\min_{u(t)} J = \int_{t_0}^{t_f} F(t, X(t), U(t), \Delta) dt \tag{6.12}$$

By assuming the continuity and analyticity of the function $F(t, X(t), U(t), \Delta)$, the objective is to determine the behavior of the extremum by utilizing the fundamental lemma of the calculus of variations. This approach ensures that the boundary conditions are satisfied.

To determine the extremum value based on fundamental calculus variation, the Lagrangian must be first obtained. Assume the following augmented function:

$$J_a = \int_{t_0}^{t_f} F\big(t, X(t), \dot{X}(t), \lambda(t), \Delta\big) dt \tag{6.13}$$

where $\lambda(t)$ is the Lagrange multiplier ($[L] = [\underline{l}, \overline{l}]$) that can be achieved as follows:

$$\begin{aligned} [L] &= L\big(t, X(t), \dot{X}(t), \lambda(t), t, \Delta\big) \\ &= F\big(t, X(t), \dot{X}(t), t, \Delta\big) + \lambda(t) G\big(t, X(t), \dot{X}(t), \Delta\big) \end{aligned} \tag{6.14}$$

If $G\big(t, X(t), \dot{X}(t), \Delta\big) = 0$, the Lagrangian equation will be as follows:

$$L\big(t, X(t), \dot{X}(t), t, \Delta\big) = F\big(t, X(t), \dot{X}(t), t, \Delta\big) \tag{6.15}$$

Note that the setting of the constraint and the augmented function are considered equivalent, i.e., $J = J_a$.

Assuming that the values are optimal, then we can apply the variation,

$$
\begin{aligned}
&x_i(t) = x_i^*(t) + \delta x_i(t), \\
&\dot{x}_i^*(t) = \dot{x}_i^*(t) + \delta \dot{x}_i(t), \\
&\Delta J_a = J_a\big(x_i^*(t), \delta x_i(t), \dot{x}_i^*(t), \delta \dot{x}_i(t), t, \Delta\big) - J_a\big(x_i^*(t), \dot{x}_i^*(t), t, \Delta\big), \\
&i = 1, 2, \dots, n
\end{aligned}
\tag{6.16}
$$

where n is the number of variables.

To continue the work and expand upon the above function, Taylor expansions have been employed. We assume the first derivatives of the functional J_a and define interval Taylor expansions as follows:

$$
F_T([x]) \triangleq ([\Delta x].\nabla)f + R_2
\tag{6.17}
$$

where R_2 presents the second-order error, and

$$
\delta J_a = \int_{t_0}^{t_f} \left\{ \left(\frac{\partial[L]}{\partial[x_i]} \right)^* . \delta x_i + \left(\frac{\partial[L]}{\partial[\dot{x}_i]} \right)^* . \delta \dot{x}_i \right\}
\tag{6.18}
$$

Note that the interval definition of partial derivation is presented in Chapter 3. By integrating part by part, the terms $\delta \dot{x}_i(t)$ can be written based on $\delta x_i(t)$ as follows:

$$
\int_{t_0}^{t_f} \left(\left(\frac{\partial[L]}{\partial[\dot{x}_i]} \right)^* . \delta \dot{x}_i(t) \right) dt = \int_{t_0}^{t_f} \left(\frac{\partial[L]}{\partial[\dot{x}_i]} \right)^* . \frac{d}{dt} (\delta(x_i(t)) dt = \left[\left(\frac{\partial[L]}{\partial[\dot{x}_i]} \right)^* \delta(x_i(t)) \right]_{t_0}^{t_f}
$$
$$
- \int_{t_0}^{t_f} \frac{d}{dt} \left(\frac{\partial[L]}{\partial[\dot{x}_i]} \right)^* . \delta(x_i(t) dt
\tag{6.19}
$$

Therefore, the first variation of the functional J_a will be

$$
\delta J_a = \int_{t_0}^{t_f} \left(\left(\frac{\partial[L]}{\partial[x_i]} \right)^* \cdot \delta(x_i(t) \right) dt + \left[\left(\frac{\partial[L]}{\partial[\dot{x}_i]} \right)^* \delta(\dot{x}_i(t)) \right]_{t_0}^{t_f} \ominus_g
$$
$$
- \int_{t_0}^{t_f} \left(\frac{d}{dt} \left(\frac{\partial[L]}{\partial[\dot{x}_i]} \right)^* \cdot \delta x_i(t) \right) dt
\tag{6.20}
$$

Boundary points are not allowed to undergo any changes due to the assumption of constant time, initial, and final states. To clarify, any modification in the boundary points is prohibited, i.e.,

$$\delta(x_i(t_0)) = \delta(x_i(t_f)) = \{0\} \tag{6.21}$$

By adding the boundary conditions to the first augmented variations,

$$\delta J_a = \int_{t_0}^{t_f} \left(\left(\frac{\partial[L]}{\partial[x_i]} \right) \ominus_g \frac{d}{dt} \left(\frac{\partial[L]}{\partial[\dot{x}_i]} \right) . \delta(x_i(t)) \right) dt = \{0\} \tag{6.22}$$

The next step is to use the principal theory. Based on the principal theory of the calculus of the variation, the first variations are considered zero, and $\lambda(t)$ is chosen arbitrarily and available so that the independent variant coefficients, $\delta x_i(t)$ are zero; in other words,

$$\left(\frac{\partial[L]}{\partial[x_i]} \right) \ominus_g \frac{d}{dt} \left(\frac{\partial[L]}{\partial[\dot{x}_i]} \right) = \{0\} \tag{6.23}$$

Finally, the necessary conditions for the functional optimization in the presence of boundary conditions according to the interval Euler–Lagrange equations are

$$\left(\frac{\partial[L]}{\partial[x_i]} \right)^* \ominus_g \frac{d}{dt} \left(\frac{\partial[L]}{\partial[\dot{x}_i]} \right)^* = \{0\}$$
$$\left(\frac{\partial[L]}{\partial\lambda} \right)^* \ominus_g \frac{d}{dt} \left(\frac{\partial[L]}{\partial\dot{\lambda}} \right)^* = \{0\} \tag{6.24}$$

Recent equations are achieved based on assuming the $\delta(x_i(t))$ independent. In addition, the interval Lagrange $[L]$ is independent of the $\dot{\lambda}(t)$. Another point is that adding $\lambda(t)$ allows us to consider the variables of the $x_i(t)$ to be independent of each other.

6.6 Analysis of the Problem of the Interval Optimal Control Based on Euler–Lagrange Equations

Consider the dynamical system of Eqs. (6.11) and (6.12). The vector (scalar) F is continuous and can be linear or nonlinear, and to have an appropriate solution, it must meet Lipschitz and the derivative conditions. Given the dynamics of the system,

$$\Phi\left(t, X(t), \dot{X}(t), U(t), \Delta\right) = F(t, X(t), U(t), \Delta) - \dot{X}(t) = \{0\} \tag{6.25}$$

Considering the Lagrangian problem, the performance index will be

$$L\left(t, X(t), \dot{X}(t), U(t), \Delta\right) = G\left(t, X(t), \dot{X}(t), U(t), \Delta\right)$$
$$+ \lambda^T(t)\left(F\left(\Phi t, X(t), \dot{X}(t), U(t), \Delta\right)\right) \tag{6.26}$$

where $f = \left[\underline{f},\overline{f}\right]$ and λ is the Lagrange coefficient. Then, to find the optimal value for the state variable $(x^*(t))$ to minimize $J(x)$, based on the Euler–Lagrange function, i.e.,

$$\left(\frac{\partial[L]}{\partial X}\right) = \frac{d}{dt}\left(\frac{\partial[L]}{\partial \dot{X}}\right)$$

$$\left(\frac{\partial[L]}{\partial U}\right) = \frac{d}{dt}\left(\frac{\partial[L]}{\partial \dot{U}}\right) \tag{6.27}$$

$$\left(\frac{\partial[L]}{\partial \lambda}\right) = \frac{d}{dt}\left(\frac{\partial[L]}{\partial \dot{\lambda}}\right)$$

In the following, using the interval analysis and the Hukuhara difference [13] to solve the above equation,

$$\left(\frac{\partial[L]}{\partial X_i}\right) = \frac{d}{dt}\left(\frac{\partial[L]}{\partial \dot{X}_i}\right)$$

$$\left(\frac{\partial[L]}{\partial U_i}\right) = \frac{d}{dt}\left(\frac{\partial[L]}{\partial \dot{U}_i}\right) \tag{6.28}$$

$$\left(\frac{\partial[L]}{\partial \lambda}\right) = \frac{d}{dt}\left(\frac{\partial[L]}{\partial \dot{\lambda}}\right)$$

Finally, the optimal control rule can be obtained by the Euler–Lagrange equations.

6.7 Solving Optimal Control Problems with Interval Uncertainties: Interval Runge–Kutta Method

Time discretization was first introduced in the 1960s for solving OCPs [14]. In that context, Polak provides an overview of previous works [14]. One of the discrete methods used for time discretization is the Runge–Kutta method (RKM). This method plays a crucial role in numerically solving differential systems. Therefore, in this section, we utilize an interval version of the RKM to solve OCPs with uncertainties.

Our proposed approach is based on a new representation of the RKM, which incorporates the Hukuhara difference. By combining these two techniques, we aim to solve OCPs with interval uncertainties. In this method, we first utilize Lagrange coefficients to satisfy the necessary conditions. Then, through algebraic manipulation, we convert them into interval differential equations. This process allows us to obtain a confidence interval that guarantees a solution for the OCPs.

Since the RKM is only applicable for solving initial value problems (IVPs), we employ the shooting method to handle boundary limitations in solving the OCPs. Now, let us consider the following linear OCP [15]:

$$
\begin{aligned}
\min_{u(t) \in \Omega = \delta_1 \times \delta_2} & J(x9t0, u(t), \Delta) \\
&= \int_0^1 \delta_1 x^2(t) + u^2(t) dt.
\end{aligned}
\tag{6.29}
$$

Subject to:

$$
\dot{x}(t) = \delta_2 u(t),
\tag{6.30}
$$

where

$$
\delta_1, \delta_2 \in \left[\frac{1}{2}, \frac{3}{2}\right], x(0) = \{1\}.
\tag{6.31}
$$

Step 1. Compute the Lagrangian function:

$$
\begin{aligned}
L(x, \dot{x}, u, \lambda, t) &= g(x, u, t) + \lambda(t)(\dot{x}(t) - f) \\
&= \delta_1 x^2(t) + u^2(t) + \lambda(t)(\dot{x}(t) - \delta_2 u(t)).
\end{aligned}
\tag{6.32}
$$

Step 2. Apply Euler–Lagrange equations to the problem:

$$
\begin{cases}
\dot{x}(t) - \delta_2 u(t) = 0, \\
2\delta_1 x(t) + \dot{\lambda} = 0, \\
2u(t) - \delta_2 \lambda = 0.
\end{cases}
\tag{6.33}
$$

Step 3. By solving the equation above over $x(t)$, the following differential equation results:

$$
\begin{cases}
\ddot{x}(t) + \gamma x(t) = 0, \\
\gamma = \delta_1 \delta_2^2.
\end{cases}
\tag{6.34}
$$

Step 4. Apply Modal Interval Arithmetic to solve the problem:
1) Compute the interval values in the problem, i.e.,

$$
\delta_1 \delta_2^2 = \left[\frac{1}{2}, \frac{3}{2}\right]^3 = \left[\frac{1}{8}, \frac{27}{8}\right],
\tag{6.35}
$$

2) Form the interval ordinary differential equation (IODE) systems:

$$\overline{x}(t): \begin{cases} \ddot{x}(t) - \dfrac{1}{8}x(t) = 0, \\ x(0) = 1. \end{cases} \qquad \underline{x}(t): \begin{cases} \ddot{x}(t) - \dfrac{27}{8}x(t) = 0, \\ x(0) = 1. \end{cases} \tag{6.36}$$

It is important to note that for simplifying the interval Runge–Kutta Methods (IRKMs) for differential equations with order greater than 1, we can first transform them into a first-order dynamic system.

$$\overline{x}(t): \begin{cases} \dot{x}_1(t) = x_2(t), \\ \dot{x}_2(t) = \dfrac{1}{8}x_1(t), \\ x(0) = 1, \end{cases} \qquad \underline{x}(t): \begin{cases} \dot{x}_1(t) = x_2(t), \\ \dot{x}_2(t) = \dfrac{27}{8}x_1(t), \\ x(0) = 1. \end{cases} \tag{6.37}$$

Figure 6.2 shows the solution of the *IRKM* for this example.

In the field of optimal control, we often come across problems that involve a fixed final state. This can be challenging because RKMs are primarily designed for solving IVPs. However, there is a popular approach known as the shooting

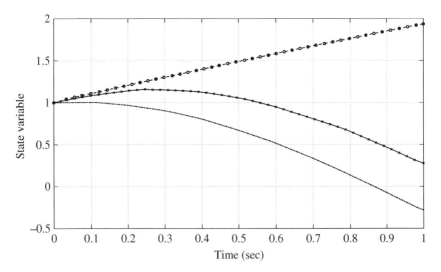

Figure 6.2 The solution of the optimal state for the proposed IRKMs on the example: (the outer intervals) IRKM^{1st}, (× and o marks) IRKM^{2nd}, and (dashed and dashed line) IRKM^{4th}.

method, which has been proposed to overcome this limitation and is described in [16, 17]. By transforming BVPs into IVPs, the shooting method offers a solution for handling OCPs with fixed final states. Below is a basic pseudocode demonstrating the single shooting method:

1) Choose $t^{(0)}$
2) Choose the step size (h) as $b\text{-}a = h^*N$, where N is the number of steps
3) For $k = 1, 2, \dots$ until convergence, **do**
4) $i = 0, y_0^1 = \alpha, y_0^2 = t^{(k)}, z_0^1 = 0, z_0^2 = 1,$
5) For $i = 0, 1, \dots, N-1,$
6) Do while $\varepsilon < |t^{(k+1)} - t^{(k)}|$,
7) Call *IRKM*
8) **End do**

The pseudocode of the proposed method is summarized as follows:

Input:

$h, T = [a, b]$

$\Delta_K = [\underline{\delta}, \overline{\delta}]$, ($K$: number of uncertainties)

$X_j = [\underline{x_j}, \overline{x_j}], j = 1, 2, \dots, m$ (m: number of states)

$F(t, X(t)),$

$$\Phi = f\left(t, X_i^n(t), \Delta\right),$$

Output:

$$X, U$$

Start: Apply the following operations to the interval OCP:

if the OCP system is IVP:

Form Lagrangian function

Apply Euler–Lagrange equations

Apply Interval arithmetic and generate ordinary differential equation (ODE) from OCP

if X is monotonic

if X is increasing

$$\text{if: } \underline{x}_i^n(t) = \overline{x}_i^n(t), \begin{cases} \underline{x}_i^{n+1}(t) = \underline{x}_i^n(t) + h\underline{\Phi}_i, \\ \overline{x}_i^{n+1}(t) = \overline{x}_i^n(t) + h\overline{\Phi}_i. \end{cases}$$

$$if: \underline{x}_i^n(t) = \overline{x}_i^n(t), \begin{cases} \underline{x}_i^{n+1}(t) = \underline{x}_i^n(t) + h\overline{\Phi}_i, \\ \overline{x}_i^{n+1}(t) = \overline{x}_i^n(t) + h\underline{\Phi}_i. \end{cases}$$

else-if Y is decreasing

$$if: \underline{x}_i^n(t) \le \overline{x}_i^n(t), \begin{cases} \underline{x}_i^{n+1}(t) = \underline{x}_i^n(t) + h\underline{\Phi}_i, \\ \overline{x}_i^{n+1}(t) = \overline{x}_i^n(t) + h\overline{\Phi}_i. \end{cases}$$

$$if: \underline{x}_i^n(t) \ge \overline{x}_i^n(t), \begin{cases} \underline{x}_i^{n+1}(t) = \underline{x}_i^n(t) + h\underline{\Phi}_i, \\ \overline{x}_i^{n+1}(t) = \overline{x}_i^n(t) + h\overline{\Phi}_i. \end{cases}$$

end-if
end-if

else-if the OCP system is BVP:
Apply shooting method
Return to **Start**
end-if

end

Wu et al. proposed an interval Euler method for solving ODEs in OCPs, which is described in detail in [18]. However, this method has limitations when it comes to solving problems governed by nonlinear differential equations.

To address these limitations and handle problems with interval uncertainties, a combination of the Runge–Kutta (RK) method and interval arithmetic can be used to improve the solution. Additionally, the Hukuhara differencing method is employed to address certain shortcomings of interval arithmetic. The RK-based methods, utilizing the shooting method, can be applied to solve linear and nonlinear systems using interval arithmetic.

By considering the approach presented in [18], designed for IVPs, along with its modified version based on the shooting method, a comprehensive comparison can be made for boundary value OCPs.

- *Numerical example*

In this section, we analyze the proposed method by applying it to several case studies. These selected case studies are intentionally kept simple to effectively demonstrate the various steps involved in the proposed *IRKM* and to facilitate result analysis.

While the case studies themselves are sourced from different references, we introduce interval uncertainties to them to evaluate the performance of the proposed IRKM. By incorporating interval uncertainties, we can better assess the effectiveness and robustness of the IRKM.

Case Study 6.1 Interval Linear Optimal Control

Consider the Following Linear Optimal Control [19]

$$\begin{matrix} \min \\ u(t) \in \Omega = \delta_1 \times \delta_2 \times \delta_3 \end{matrix} J(X(t), u(t), \Delta) = \int_0^1 \delta_1 x^2(t) + \delta_2 u^2(t) dt, \qquad (6.38)$$

subject to:

$$\dot{x}(t) = \delta_3 x(t) + u(t), \qquad (6.39)$$

where

$$\delta_1 \in [1,3], \delta_2 \in \left[\frac{1}{8}, \frac{3}{8}\right], \delta_3 \in \left[\frac{1}{2}, \frac{3}{2}\right],$$
$$x(0) = \{1\}, x(1) = [1, 1.2]. \qquad (6.40)$$

The system has interval uncertainties in both boundary conditions and index performance.

Step 1. Compute the Lagrangian function:

$$L(x, \dot{x}, u, \lambda, t) = \delta_1 x^2(t) + \delta_1 u^2(t) + \lambda(t)(\dot{x}(t) - \delta_3 x(t) - u(t)). \qquad (6.41)$$

Step 2. Apply Euler–Lagrange equations to the problem:

$$\begin{cases} \dot{x}(t) - \delta_3 x(t) - u(t) = 0, \\ 2\delta_2 u(t) - \lambda = 0, \\ 2\delta_1 x(t) - \delta_3 \lambda - \dot{\lambda} = 0. \end{cases} \qquad (6.42)$$

Step 3. By solving the equation above over $x(t)$, the following differential equation results:

$$\begin{cases} \ddot{x}(t) - \gamma x(t) = 0, \\ \gamma = \dfrac{\delta_1}{\delta_2} + \delta_3^2, \end{cases} \qquad (6.43)$$

Step 4. Apply Modal Interval Arithmetic to solve the problem:
3) Compute the interval values in the problem, i.e.,

$$\left. \begin{matrix} \dfrac{\delta_1}{\delta_2} = \delta_1 \times \dfrac{1}{Dual(\delta_2)} = \left\{\dfrac{3}{8}\right\}, \\ \delta_3^2 = \left[\dfrac{1}{4}, \dfrac{9}{4}\right], \end{matrix} \right\} \rightarrow \gamma = \dfrac{\delta_1}{\delta_2} + \delta_3^2 = \left[\dfrac{5}{8}, \dfrac{21}{8}\right]. \qquad (6.44)$$

4) Form the interval ODE systems:

$$\overline{x}(t): \begin{cases} \ddot{x}(t) - \dfrac{5}{8}x(t) = 0, \\ x(0) = 1, x(1) = 1. \end{cases} \qquad \underline{x}(t): \begin{cases} \ddot{x}(t) - \dfrac{21}{8}x(t) = 0, \\ x(0) = 1, x(1) = 1.2. \end{cases} \qquad (6.45)$$

It is important to note that for simplifying the IRKMs for differential equations with order greater than 1, we should first transform them into a first-order dynamic system:

$$\overline{x}(t): \begin{cases} \dot{x}_1(t) = x_2(t), \\ \dot{x}_2(t) = \dfrac{5}{8}x_1(t), \\ x(0) = 1, \\ x(1) = 1. \end{cases} \qquad \underline{x}(t): \begin{cases} \dot{x}_1(t) = x_2(t), \\ \dot{x}_2(t) = \dfrac{21}{8}x_1(t), \\ x(0) = 1, \\ x(1) = 1.2. \end{cases} \qquad (6.46)$$

Solving the IODE by the proposed IRKM results:

Figure 6.3 illustrates that as the order of the IRKM increases, the computed results gradually approach the actual solution. This observation indicates that

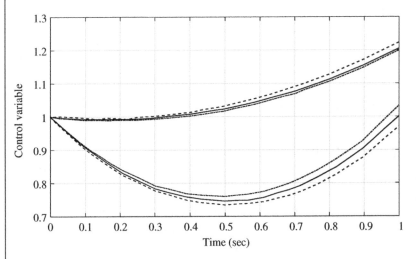

Figure 6.3 The solution of the proposed IRKMs for Case Study 6.1: (lower line and upper dot-dashed) Exact, (lower dot-dashed and upper dashed) IRKM^{1st}, (lower dashed, upper line) IRKM^{2nd}, and (lower line and upper dot-dashed) IRKM^{4th} for optimal control.

higher-order IRKMs offer greater accuracy in approximating solutions to a given problem. Consequently, higher-order methods exhibit improved numerical stability and convergence rates. Moreover, as the order of the IRKM increases, the computational effort required to obtain a solution decreases. Higher-order methods typically require fewer time steps to achieve a desired level of accuracy compared to lower-order methods.

To conduct a comprehensive analysis and comparison of the interval Runge–Kutta methods, the obtained results are presented in Table 6.1. To evaluate the accuracy of the computed solutions, an interval L2 norm is calculated and utilized to compare the performance of different IRKMs. The interval L2 norm is a mathematical formula employed to assess the distance between two interval integers, denoted as X and Y. It is expressed as follows:

$$\|X - Y\|_2 = \sqrt{\left(\underline{x} - \underline{y}\right)^2 + (\overline{x} - \overline{y})^2} \tag{6.47}$$

where $\underline{x}$ and $\underline{y}$ are vectors.

Indeed, Table 6.1 provides a comprehensive summary of the computed results obtained using different orders of the IRKM. It is demonstrated in the table that as the order of the IRKM increases, the error in the computed solution decreases. This finding emphasizes the significance of selecting a higher-order IRKM when solving a given problem, as it can greatly enhance the accuracy of the numerical solution.

Table 6.1 Comparison of the IRKMs and exact values for optimal control ($u^*(t)$).

Time	Exact Method Lower bound	Exact Method Upper bound	IRKM1st [18] Lower bound	IRKM1st [18] Upper bound	IRKM2nd Lower bound	IRKM2nd Upper bound	IRKM4th Lower bound	IRKM4th Upper bound
0.0	0.72	1.22	0.64	1.15	0.73	1.20	0.72	1.20
0.2	0.96	1.55	0.87	1.46	0.97	1.52	0.97	1.52
0.4	1.42	2.11	1.31	1.99	1.43	2.07	1.44	2.07
0.6	2.55	3.30	2.38	3.13	2.56	3.23	2.56	3.24
0.8	6.25	6.61	6.36	6.00	6.27	6.49	6.24	6.47
1.0	17.95	20.51	17.71	20.33	17.61	20.42	17.81	20.4
$\|Error\|_2$	—		1.5626		0.6535		0.4575	

Furthermore, the table enables a direct comparison of the performance of various IRKMs, allowing for the identification of the most suitable method based on the desired level of accuracy and computational efficiency for a specific application. In this case, the hybrid IRKM[4th] and IRKM[5th] methods are used to simulate the exact method, indicating their potential in achieving accurate solutions while minimizing computational effort.

Case Study 6.2   Nonlinear optimal control

Consider the following nonlinear optimal control as follows [19]

$$\min_{u(t)\,\in\,\Omega\,=\,\delta_1\,\times\,\delta_2} J(x(t),u(t),\Delta) = \int_0^1 \delta_1 x^2(t) + \delta_2 u^2(t)dt, \tag{6.48}$$

subject to:

$$\dot{x}(t) = \delta_2 x^2(t) + u(t), \tag{6.49}$$

where

$$\delta_1 \in [1,2], \delta_2 \in \left[\frac{3}{2},\frac{5}{2}\right],$$

$$x(0) = \{-1\}, x(1) = \{0.5\}$$

Here, the system conditions are degenerate intervals, but the performance index has interval uncertainties.

After performing the initial operations, the interval ODE systems are achieved as follows:

$$\underline{x}(t): \begin{cases} \dot{x}_1(t) = x_2(t), \\ \dot{x}_2(t) = \dfrac{25}{8}x_1^3(t), \\ x(0) = -1, \\ x(1) = 0.5. \end{cases} \qquad \underline{x}(t): \begin{cases} \dot{x}_1(t) = x_2(t), \\ \dot{x}_2(t) = \dfrac{25}{8}x_1^3(t), \\ x(0) = -1, \\ x(1) = 0.5. \end{cases} \tag{6.50}$$

By applying the proposed IRKM to solve the IODE, the solution is obtained as follows:

In [20], an interval Adomian method based on our technique is introduced for comparison alongside the IRKMs. The computed solutions for both the state and control variables are illustrated in Figure 6.4a–c, respectively. The results

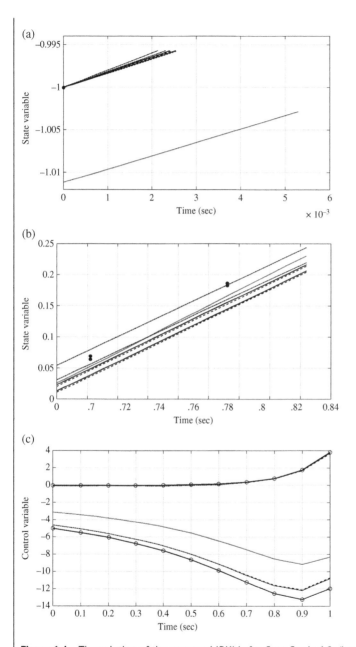

Figure 6.4 The solution of the proposed IRKMs for Case Study 6.2: (lower and upper line-circle) Exact, (lower and upper small dotted line) IRKM^{1st}, (lower and upper dashed line) IRKM^{2nd}, and (lower and upper dotted-dashed line) IRKM^{4th} for (a) and (b) optimal state and (c) optimal control.

indicate that while the Adomian method solves with a relatively small error, it does not perform as effectively as the IRKMs in the initial stages of the computation. As the computation progresses, the Adomian method yields wider intervals, which may not be desirable in interval computations. This suggests that the IRKMs outperform the Adomian method in terms of accuracy and computational efficiency, particularly during the early stages of the computation. Overall, the results highlight the superiority of the IRKMs over the interval Adomian method, emphasizing their advantages in achieving higher accuracy and computational efficiency.

Table 6.2 presents the results of the analysis, focusing on two key features that need to be examined. First, it is observed that for the chosen error metric, lower-order IRKMs achieve better results. This observation can be attributed to higher-order methods being more sensitive to small fluctuations in the solution, leading to increased errors. Therefore, in terms of minimizing

Table 6.2 Comparison of the IRKMs and exact values for optimal state $(x^*(t))$.

Time	Exact method Lower bound	Exact method Upper bound	IRKM1st [18] Lower bound	IRKM1st [18] Upper bound	IRKM2nd Lower bound	IRKM2nd Upper bound	IRKM4th Lower bound	IRKM4th Upper bound	Adomian method [20] Lower bound	Adomian method [20] Upper bound
0.0	−1.00	−1.00	−1.00	−1.00	−1.00	−1.00	−1.00	−1.00	−1.011	−1.00
0.1	−0.829	−0.82	−0.83	−0.82	−0.83	−0.82	−0.83	−0.82	−0.85	−0.80
0.2	−0.67	−0.66	−0.68	−0.67	−0.67	−0.67	−0.65	−0.67	−0.70	−0.63
0.3	−0.52	−0.51	−0.53	−0.52	−0.53	−0.52	−0.50	−0.51	−0.55	−0.48
0.4	−0.37	−0.36	−0.39	−0.38	−0.38	−0.37	−0.36	−0.37	−0.40	−0.34
0.5	−0.23	−0.22	−0.25	−0.23	−0.24	−0.23	−0.22	−0.11	−0.25	−0.11
0.6	−0.08	−0.07	−0.10	−0.09	−0.10	−0.09	−0.08	−0.07	−0.06	−0.06
0.7	0.062	0.066	0.034	0.044	0.0351	0.045	0.053	0.077	0.053	0.077
0.8	0.207	0.210	0.192	0.215	0.176	0.186	0.175	0.185	0.200	0.185
0.9	0.353	0.354	0.317	0.327	0.318	0.328	0.330	0.354	0.350	0.354
1.0	0.5	0.5	0.459	0.469	0.460	0.469	0.470	0.51	0.500	0.492
$\|Error\|_2$	—		0.2665		0.2747		0.2913		0.3562	

the error metric, it is preferable to use IRKMs with lower orders. Second, the tableillustrates that using higher-order IRKMs is advantageous in ensuring that the computed solution falls within the desired confidence interval. Higher-order methods can provide a more accurate approximation of the true solution, reducing the likelihood of the computed solution lying outside the specified confidence interval. Taken together, the table provides valuable insights into the performance of different IRKMs. It aids in the selection of the most suitable method for a given application, considering both the desired level of accuracy and the requirement of maintaining a specific confidence interval for the computed solution.

Case Study 6.3   Interval OCP Tracking System

Consider the Following Tracking OCP Where the Dynamical System is Nonlinear and the Performance Index is Quadric [21]

$$\min_{u(t)\in\Omega \,=\, \delta_1 \,\times\, \delta_2} J(x(t), u(t), \Delta) = \int_0^1 (\delta_1 - x(t))^2 + u^2(t)\, dt, \tag{6.51}$$

subject to:

$$\dot{x}(t) = -\delta_2 \sqrt{x(t)} + u(t), \tag{6.52}$$

where

$$\delta_1 \in [1, 3], \delta_2 \in [-1, -0.25],$$

$$x(0) = \{0\}, x(1) = \left[\frac{1}{2}, \frac{3}{2}\right]. \tag{6.53}$$

Figure 6.5 displays the tracking system.

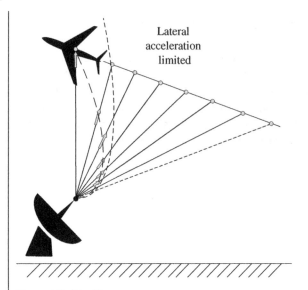

Figure 6.5 Tracking system.

After performing the initial operations, the interval ODE systems are obtained as follows:

$$\overline{x}(t): \begin{cases} \dot{x}_1(t) = x_2(t), \\ \dot{x}_2(t) = x(t) - \dfrac{191}{64}x_1(t), \\ x(0) = 0, \\ x(1) = 3. \end{cases} \qquad \underline{x}(t): \begin{cases} \dot{x}_1(t) = x_2(t), \\ \dot{x}_2(t) = x(t) - \dfrac{3}{4}x_1(t), \\ x(0) = 0, \\ x(1) = 1. \end{cases} \qquad (6.54)$$

Solving the IODE by the proposed IRKM results in the following solution:

From Figure 6.6, various orders of IRKMs are applied to the system and compared with the exact values. It can be observed that by increasing the order of IRKM, the interval bound is improved. The proposed method is an extension of the RKM based on interval arithmetic, utilizing interval analysis theory and its extension approaches, including the generalized Hukuhara method, to enhance its performance. The proposed method introduces a new description called forward representation. Its primary objective is to determine the interval optimal control and state vector of OCPs with interval uncertainties through a direct solution method based on an improved version of the RKM. Numerical examples demonstrate that one of the advantages of the proposed interval method over the Euler method is its efficiency, particularly when dealing with a large range and number of uncertain variables.

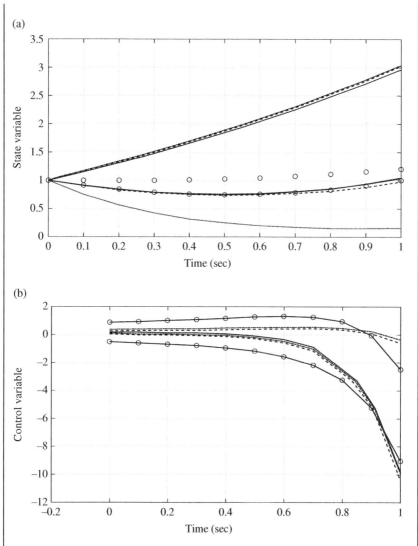

Figure 6.6 The solution of the proposed IRKMs for Case Study 6.1: (circle) Exact, (line) IRKM^{1st}, (small-dotted) IRKM^{2nd}, and (small-dashed) IRKM^{4th} for (a) and (b) optimal control.

6.8 Optimal Control of Problems with Interval Uncertainties Using the Chebyshev Inclusion Method

In this section, we present the interval Chebyshev method, which was introduced in Chapter 5. Initially, the interval model is employed to capture the uncertainties associated with the desired optimal control. These uncertainties can be represented as intervals that encompass the initial conditions of the system or even its parameters.

Let us consider a dynamical system model with interval uncertainties, described by the following ODE:

$$\dot{x}(t) = f(t, x(t), u(t), \Delta), t \in (t_0, t_f),\tag{6.55}$$

And the boundary conditions are

$$x(t_0) = X_0, x(t_f) = X_f\tag{6.56}$$

where $u(t)$ and $x(t)$ are control and state variables, respectively, $\Delta = [\delta_1, \delta_2, ..., \delta_n]$ describes the system uncertainties which are bounded in an interval, and X_0 and X_f are the initial and the final states of the system, respectively.

Consider the constrained minimization problem as follows:

$$\min_{u(t)} J(x(t), u(t), \Delta) = \int_{t_0}^{t_f} G(t, x(t), u(t), \Delta) dt,\tag{6.57}$$

In this case, $J(x(t), u(t), \Delta) = \left[\underline{j}(x(t), u(t), \Delta), \bar{j}(x(t), u(t), \Delta) \right]$ describes the interval-valued performance index.

The vector (scalar) function G can be linear or nonlinear and is assumed to be continuous, Lipchitz, and differentiable concerning their arguments. If $t_0 \neq -1$ or $t_f \neq 1$, then for employing the proposed Chebyshev polynomials, the time transformation should be performed as follows:

$$\tau = \frac{t_f - t_0}{2} t + \frac{t_f + t_0}{2},\tag{6.58}$$

By replacing Eq. (6.55) with Eq. (6.57),

$$u(t) = f\left(\frac{t_f - t_0}{2} t + \frac{t_f + t_0}{2}, x(t), \dot{x}(t), \Delta \right),\tag{6.59}$$

Therefore, the corresponding initial conditions of the trajectory $x(t)$ can be described as follows:

$$x(-1) = x_0, x(1) = x_f \tag{6.60}$$

And the performance index can be defined by

$$J(t) = \frac{t_1 - t_0}{2} \int_{-1}^{1} G\left(\frac{t_f - t_0}{2}t + \frac{t_f + t_0}{2}, x(t), \dot{x}(t), \Delta\right) dt, \tag{6.61}$$

An indirect method is here employed for solving the OCPs. From Eq. (6.55), the function is defined as

$$\phi(t, x(t), \dot{x}(t), u(t), \Delta) = f(t, x(t), u(t), \Delta) - \dot{x}(t) = 0, \tag{6.62}$$

By considering the Lagrange problem, the functional criteria can be described as

$$\begin{aligned} L(x(t), \dot{x}(t), u(t), \lambda, t, \Delta) &= G(t, x(t), \dot{x}(t), u(t), \Delta) \\ &+ \lambda^T(t)(\phi(t, x(t), \dot{x}(t), u(t), \Delta)), \end{aligned} \tag{6.63}$$

where $G = [\underline{G}, \overline{G}]$ and λ is the Lagrange multiplier. Afterward, to find the x^* for minimizing $J(x)$ (according to the Euler–Lagrange function),

$$\begin{aligned} \frac{\partial f}{\partial x} &= \frac{d}{dt}\left(\frac{\partial G}{\partial \dot{x}}\right), \\ \frac{\partial f}{\partial u} &= \frac{d}{dt}\left(\frac{\partial G}{\partial \dot{u}}\right), \\ \frac{\partial f}{\partial \lambda} &= \frac{d}{dt}\left(\frac{\partial G}{\partial \dot{\lambda}}\right). \end{aligned} \tag{6.64}$$

By using the Hukuhara derivation for solving the interval equation above [22],

$$\begin{aligned} \frac{\partial f}{\partial x} &= \left[\min\left\{\frac{d}{dt}\left(\frac{\partial \underline{G}}{\partial \dot{x}}\right), \frac{d}{dt}\left(\frac{\partial \overline{G}}{\partial \dot{x}}\right)\right\}, \max\left\{\frac{d}{dt}\left(\frac{\partial \underline{G}}{\partial \dot{x}}\right), \frac{d}{dt}\left(\frac{\partial \overline{G}}{\partial \dot{x}}\right)\right\}\right], \\ \frac{\partial f}{\partial u} &= \left[\min\left\{\frac{d}{dt}\left(\frac{\partial \underline{G}}{\partial \dot{u}}\right), \frac{d}{dt}\left(\frac{\partial \overline{G}}{\partial \dot{u}}\right)\right\}, \max\left\{\frac{d}{dt}\left(\frac{\partial \underline{G}}{\partial \dot{u}}\right), \frac{d}{dt}\left(\frac{\partial \overline{G}}{\partial \dot{u}}\right)\right\}\right], \\ \frac{\partial f}{\partial \lambda} &= \left[\min\left\{\frac{d}{dt}\left(\frac{\partial \underline{G}}{\partial \dot{\lambda}}\right), \frac{d}{dt}\left(\frac{\partial \overline{G}}{\partial \dot{\lambda}}\right)\right\}, \max\left\{\frac{d}{dt}\left(\frac{\partial \underline{G}}{\partial \dot{\lambda}}\right), \frac{d}{dt}\left(\frac{\partial \overline{G}}{\partial \dot{\lambda}}\right)\right\}\right], \end{aligned} \tag{6.65}$$

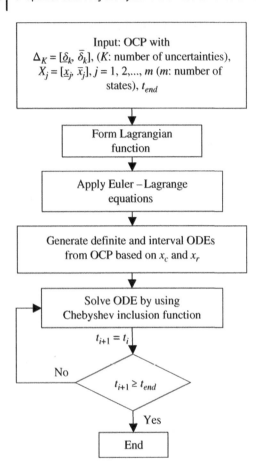

Input: OCP with
$\Delta_K = [\underline{\delta_k}, \overline{\delta_k}]$, (*K*: number of uncertainties),
$X_j = [\underline{x_j}, \overline{x_j}]$, $j = 1, 2,..., m$ (*m*: number of states), t_{end}

Form Lagrangian function

Apply Euler – Lagrange equations

Generate definite and interval ODEs from OCP based on x_c and x_r

Solve ODE by using Chebyshev inclusion function

$t_{i+1} = t_i$

No

$t_{i+1} \geq t_{end}$

Yes

End

Figure 6.7 Flowchart diagram of the proposed Chebyshev inclusion method for solving OCPs with interval uncertainties.

Finally, the optimal control laws can be found by the Euler–Lagrange equations [23]. The flowchart of the proposed method is shown in Figure 6.7.

- *Numerical examples*

In this section, our focus is on presenting and analyzing two classes of interval-valued OCPs. These case studies have been selected from various references, but we have introduced interval uncertainties to enable their analysis as problems with inherent uncertainties. The incorporation of these uncertainties allows us to assess the robustness and reliability of the obtained solutions, as well as their sensitivity to variations in the underlying parameters. By conducting a comprehensive analysis of these problems, the aim is to offer valuable insights into the characteristics of interval-valued OCPs and evaluate the effectiveness of different solution methods.

Case Study 6.4 Interval Nonlinear Optimal Control

Consider the following nonlinear optimal control where the performance index and system dynamic are considered as the interval parameters with uncertainties $\delta_1 \epsilon [1,2], \delta_2 \epsilon \left[\frac{3}{2}, \frac{5}{2}\right]$ that is expressed as follows [19]:

$$\min_{u(t)\epsilon \delta_1 \times \delta_2} J(x(t), u(t), \Delta) = \int_{t_0}^{t_f} \delta_1 u^2(t) dt, \tag{6.66}$$

subject to:

$$\dot{x}(t) = \delta_2 x^2(t) + u(t),$$
$$x(0) = -1, x(1) = 0.5. \tag{6.67}$$

The first step is to compute the Lagrangian function of the OCP as follows:

$$L(x(t), \dot{x}(t), u(t), \lambda, t, \Delta) = \delta_1 u^2(t) + \lambda(t) \left(\dot{x}(t) - \delta_2 x^2(t) - u(t)\right) \tag{6.68}$$

The next step is to apply Euler–Lagrange equations to the problem:

$$\begin{cases} 2 \, \delta_1 u(t) - \lambda(t) = 0, \\ -2\delta_2 \lambda(t) x(t) - \dot{\lambda}(t) = 0, \\ \dot{x}(t) - \delta_2 \, x^2(t) - u(t). \end{cases} \tag{6.69}$$

By solving the equation above over $x(t)$,

$$-4\delta_1\delta_2 \times x(t)\dot{x}(t) + 2\delta_2^2 \times x^3(t) - 2\delta_1\ddot{x}(t) + 4 - 4\delta_1\delta_2 \times x(t)\dot{x}(t) = 0,$$
$$\longrightarrow \begin{cases} x(t) - \gamma \ddot{x}^3(t) = 0, \\ \gamma = \delta_2^2 / \delta_1. \end{cases} \tag{6.70}$$

Once the ODE is obtained from the OCP, we apply the Chebyshev inclusion method and the second-order interval Taylor method to solve the ODE problem over the time interval of 0–1 second. For the Chebyshev inclusion function, we select $k = 4$. Both the interval solutions obtained from the Taylor method and the Chebyshev method tightly enclose the scanning interval during the initial stages. However, the Chebyshev inclusion method exhibits a slightly tighter bound compared to the interval Taylor method. The region between the two black lines represents the outcome of the proposed interval Chebyshev method, while the red lines represent the interval Taylor solution for the considered ODE.

Case Study 6.5   An Example of the Interval Optimal Tracking Control

In many real-world scenarios, the objective of optimal control is to track and guide the behavior of a specific system as it evolves. This type of problem is referred to as tracking optimal control, and it is encountered in a wide range of applications. A typical example involves a nonlinear dynamical system and a quadratic performance index [21].

$$\min_{u(t)\epsilon\delta_1 \times \delta_2} J(x(t),u(t),\Delta) = \int_{t_0}^{t_f} (\delta_1 - x(t))^2 + u^2(t)dt, \qquad (6.71)$$

subject to:

$$\dot{x}(t) = -\delta_2\sqrt{x(t)} + u(t)x(t), \qquad (6.72)$$

The objective of tracking optimal control is to design a control strategy that enables the system to follow a desired trajectory or reference signal while minimizing the deviation between the actual and desired trajectories. This task can be particularly challenging when dealing with complex, nonlinear systems that exhibit highly nonlinear behavior or other sources of uncertainty.

To solve this type of tracking OCP, various techniques and algorithms can be employed, including both indirect and direct methods. Indirect methods involve deriving necessary conditions for optimality, such as the Pontryagin maximum principle. On the other hand, direct methods rely on the discretization or parameterization of the control and state variables.

In the given scenario, both the target for tracking and the dynamics of the system exhibit interval uncertainties, denoted as (δ_1, δ_2).

$$\delta_1 \epsilon [1,3], \delta_2 \epsilon [-1,-0.25],$$
$$x(0) = \{0\}, x(1) = \{1\}. \qquad (6.73)$$

Centered coefficients of uncertainties are $\delta_{1c} = 2$, $\delta_{2c} = -0.375$, $\delta_{1r} = 1$, $\delta_{2r} = 0.37$. By computing the Lagrangian function:

$$L(x(t),\dot{x}(t),u(t),\lambda,t,\Delta) = (\delta_1 - x(t))^2$$
$$+ u^2(t) + \lambda(t)\left(\dot{x}(t) + \delta_2\sqrt{x(t)} - u(t)\right) \qquad (6.74)$$

Using Euler–Lagrange equations to the problem results:

$$\begin{cases} 2\ u(t) - \lambda(t) = 0, \\ -2(\delta_1 - x(t)) + \dfrac{\delta_2}{2\sqrt{x(t)}}\lambda(t) - \dot{\lambda}(t) = 0, \\ \dot{x}(t) + \delta_2\sqrt{x(t)} - u(t). \end{cases} \qquad (6.75)$$

By applying the Lagrangian function and Euler–Lagrange equations, the ODE above over $x(t)$ is

$$
\begin{cases}
\ddot{x}(t) - x(t) + \gamma = 0, \\
\gamma = -\dfrac{1}{4}\delta_2^2 + \delta_1.
\end{cases}
\tag{6.76}
$$

This system is also solved for 1 second using the Chebyshev inclusion method and second-order interval Taylor method. The Chebyshev method with the fourth-order polynomials is utilized. The state variable and control variable results are shown in Figure 6.8 a,b, respectively. The results show that the proposed Chebyshev method can achieve tighter results and can handle the overestimation better than the Taylor method according to the wrapping effect of interval computations.

(a)

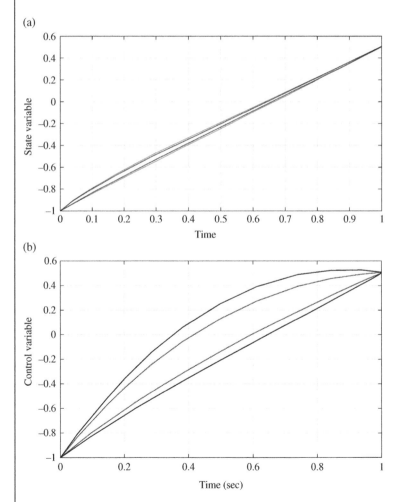

(b)

Figure 6.8 The result of interval methodology for (a) Case Study 6.1 and (b) Case Study 6.2: Interval of Taylor method (line), interval of Chebyshev inclusion function (dotted).

6.9 Piecewise Interval Chebyshev Method for OCPs

As it is maintained, the Chebyshev inclusion function is useful to reduce interval overestimation. In this part, a new method is proposed for increasing the speed and decreasing the overestimation of the inclusion Chebyshev function. Generally, a higher order of Chebyshev polynomial is required for achieving an accurate solution for an interval function. This reason increases the method's time cost. In this study, we first subdivide time interval into some subintervals and then, each period is solved by the shifted Chebyshev inclusion method as follows.

As we know, Chebyshev polynomials are defined in the interval $[-1, 1]$. For instance, if the interval has been subdivided into two subintervals ($N = 2$), we have two periods: $[-1, 0]$ and $[0, 1]$. For $N > 0$, we have

$$[-1, -1 + d], [-1 + d, -1 + 2d], ..., [-1 + (N-1)d, 1],$$
$$d = \frac{2}{N}. \tag{6.77}$$

For solving each of the above periods, a transformation should be performed to take the periods into the interval $[-1, 1]$. For instance, consider a subinterval $t = [\underline{t}, \overline{t}]$. This can be transformed to the Chebyshev interval by the following formula:

$$\tau = \frac{\overline{t} + \underline{t}}{2} + \frac{\overline{t} - \underline{t}}{2}[-1, 1], \tag{6.78}$$

where τ describes the shifted interval.

One of the profits of this method is that the operation for each period is performed by a lower-order Chebyshev polynomial (in this paper, we consider third order).

For instance, consider the function $F(X) = \arctan(X)$, where $X \subset [-1, 1]$. By employing fifth-order Chebyshev polynomials to approximate this function, Eqs. (6.72–6.75) can be utilized to achieve the coefficients of the Chebyshev series. Equations (6.76) and (6.77) can be utilized to achieve the piecewise interval Chebyshev polynomial; by dividing the interval into three and five subintervals and applying the shifted Chebyshev polynomial, the achieved results are shown in Figure 6.9.

The main function of this problem is subdivided into three and five parts (periods). For the piecewise method, a fifth-order Chebyshev polynomial is utilized

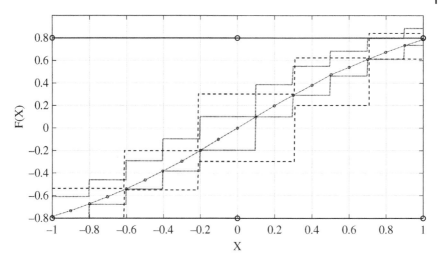

Figure 6.9 Interval approximation of arctan(x) using five-order Chebyshev polynomial (dotted-line), Piecewise 3-part (circle dot-dashed), 5-part (dashed), and 10-part (dotted) Chebyshev polynomials.

here and it should be considered that each period needs a separate Chebyshev inclusion polynomial and its coefficients. The proposed method has a lower overestimation than the method proposed by Wu et al. [24].

The proposed method in this study is based on indirect optimal control. In the proposed method, the equation, boundary conditions, and performance index are first changed into some algebraic equations. This task converts the OCP into an ODE, which can be solved then easily by the proposed interval method. Interval-valued OCP by the proposed method can be summarized by the following pseudocode:

Input:

N (number of subintervals),

$\Delta_K = \left[\underline{\delta}_k, \overline{\delta}_k\right]$, ($K$: number of uncertainties)

$X_j = \left[\underline{x}_j, \overline{x}_j\right], j = 1, 2, ..., m$ (m: number of states)

Output:

$$X, U$$

Start: Apply the following operations to the centered value of the uncertainties for the interval OCP:

Form Lagrangian function

Apply Euler–Lagrange equations

Generate definite and interval ODEs from OCP

Solve the ODEs by Chebyshev first kind and interval Chebyshev methods such as

If the interval of the ODE is in the interval [−1, 1]:

Divide the main interval into N subintervals:

If $N = 2$

Subintervals: [−1, 0] and [0, 1]

Else

$$[-1, -1 + d], [-1 + d, -1 + 2d], ..., [-1 + (N-1)d, 1], d = \frac{3}{N+1},$$

End

Apply Interval piecewise Chebyshev polynomial on each of the subintervals by

$$F(\Omega) \doteq \frac{1}{2} f_0 + [-1, 1] \left(\sum_{i=1}^{N} |f_i| \right),$$

Else if the interval is not [-1, 1]:

Transform the interval to the [-1, 1] by:

$$t = \frac{\bar{t} + \underline{t}}{2} + \frac{\bar{t} - \underline{t}}{2} [-1, 1],$$

Return

End if

End

• *Numerical example*

In this section, two classes of interval-valued OCPs are presented and analyzed. The considered case studies are from different references; however, interval uncertainties are added to these problems to analyze them as problems with uncertainties.

Case Study 6.6   Interval Nonlinear Optimal Control

Consider the Following Nonlinear Optimal Control Where the Performance Index and System Dynamic Coefficients have Interval Uncertainties [25]

$$\min_{u(t) \, \in \, \Omega \, = \, \delta_1 \, \times \, \delta_2} J(u(t)) = \int_0^1 \delta_1 u^2(t) \; dt, \tag{6.79}$$

subject to:

$$\dot{x}(t) = \delta_2 x^2(t) + u(t), \tag{6.80}$$

where δ_1 and δ_2 are system uncertainties and their values are

$$\delta_1 \in [1,2], \delta_2 \in \left[\frac{3}{2},\frac{5}{2}\right], x(0) = -1, x(1) = 0.5. \tag{6.81}$$

Here, the system conditions are degenerate intervals, but the performance index has interval uncertainties. Since we have

Step 1. Compute the Lagrangian function for centered coefficients:

$$L(x,\dot{x},u,\lambda,t) = \delta_1 u^2(t) + \lambda(t)\big(\dot{x}(t) - \delta_2 x^2(t) - u(t)\big), \tag{6.82}$$

Step 2. Apply Euler–Lagrange equations to the problem:

$$\begin{cases} 2\delta_1 u(t) - \lambda(t) = 0, \\ -2\delta_2\lambda(t)x(t) - \dot{\lambda}(t) = 0, \\ \dot{x}(t) - \delta_2 x^2(t) - u(t) = 0. \end{cases} \tag{6.83}$$

Step 3. By solving the equation above over $x(t)$,

$$-4\delta_1\delta_2 \times x(t)\dot{x}(t) + 2\delta_2^2 \times x^3(t) - 2\delta_1\ddot{x}(t) + 4 - 4\delta_1\delta_2 \times x(t)\dot{x}(t) = 0,$$
$$\rightarrow \begin{cases} \ddot{x}(t) - \gamma x^3(t) = 0, \\ \gamma = \delta_2^2/\delta_1. \end{cases} \tag{6.84}$$

Step 4. By solving the achieved ODE by the Chebyshev method and interval ODE by the interval Chebyshev method with n subdivisions (here $n = 6$), the computational results are achieved as in Figure 6.10.

The area between two black and piecewise lines is the result of the proposed interval Chebyshev method, while the colored lines show the ODE solution for the exact solution (centered values of the uncertain parameters) of the problem with six subsections (periods) with shifted piecewise Chebyshev inclusion method.

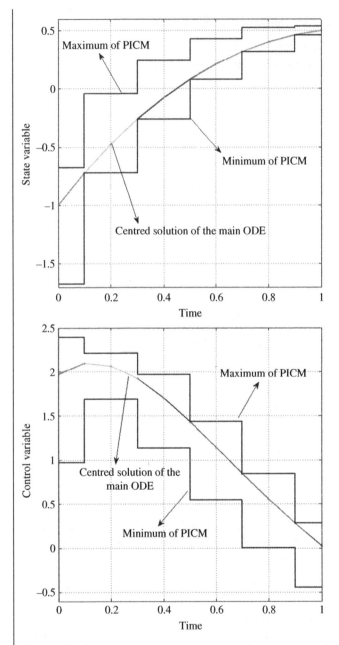

Figure 6.10 The result of interval piecewise arithmetic: centered (colored function) and applied proposed piecewise Chebyshev polynomial (black lines).

Case Study 6.7   Interval Nonlinear Tracking OCP with Quadric Performance Index

Consider the Following Tracking OCP Where the Dynamical System is Nonlinear and the Performance Index is Quadric [21]

$$
\min_{u(t) \,\in\, \Omega \,=\, \delta_1 \,\times\, \delta_2} J(u(t)) = \int_0^1 (\delta_1 - x(t))^2 + u^2(t)dt, \tag{6.85}
$$

subject to:

$$
\dot{x}(t) = -\delta_2 \sqrt{x(t)} + ux(t). \tag{6.86}
$$

Here, the target for tracking and the system dynamic has interval uncertainties (δ_1, δ_2) and

$$
\begin{aligned}
&\delta_1 \in [1,3], \delta_2 \in [-1, -0.25], \\
&x(0) = \{0\}, x(1) = \{1\}.
\end{aligned} \tag{6.87}
$$

By computing the Lagrangian function:

$$
\begin{aligned}
L(x, \dot{x}, u, \lambda, t) &= (\delta_1 - x(t))^2 + u^2(t) \\
&\quad + \lambda(t)\left(\dot{x}(t) + \delta_2\sqrt{x(t)} - u(t)\right).
\end{aligned} \tag{6.88}
$$

Using Euler–Lagrange equations to the problem results:

$$
\begin{cases}
2u(t) - \lambda(t) = 0, \\
-2(\delta_1 - x(t)) + \dfrac{\delta_2}{2\sqrt{x(t)}}\left(2\dot{x}(t) + b\sqrt{x(t)}\right) = 0, \\
\dot{x}(t) + \delta_2\sqrt{x(t)} - u(t) = 0.
\end{cases} \tag{6.89}
$$

By solving the equation above over $x(t)$,

$$
\begin{cases}
\ddot{x}(t) - x(t) + \gamma = 0, \\
\gamma = -\dfrac{1}{4}\delta_2^2 + \delta_1.
\end{cases} \tag{6.90}
$$

By assuming $n = 1000$ and solving the above ODE:

Figure 6.11 shows the proposed methods interval. As we explained before, the piecewise interval Chebyshev method (PICM) can increase the speed of the solution and decrease the interval overestimation. The speed comparison for continuous and piecewise Chebyshev inclusion methods is illustrated in Table 6.3.

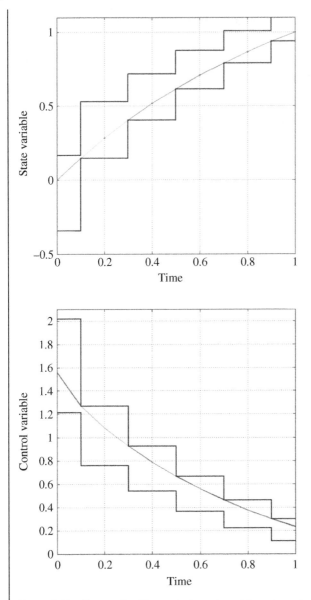

Figure 6.11 The result of interval piecewise arithmetic: centered (colored function) and applied proposed piecewise Chebyshev polynomial (black lines) for Case Study 6.7.

Table 6.3 The proposed method.

Method		Proposed Chebyshev inclusion method	Chebyshev inclusion method
Running time (sec.)	Case Study 6.6	0.84	2.6
	Case Study 6.7	1.03	4.23

It is noted that the proposed piecewise interval method can eliminate the overestimation better than the continuous method and can control the wrapping effect and the overestimation in a reasonable range.

6.10 Solving Quadratic Optimal Control Problems with Interval Uncertainties Based on Indirect Method: Interval Quadratic Regulator

In this section, a novel approach is introduced for optimal control of quadratic systems in the presence of interval uncertainty. Interval uncertainty can be viewed as an approximation of a nonlinear system, where the system is represented as a linear system with an added interval that encompasses the tolerance for error.

This equivalence allows us to extend linear methods to the realm of interval equations, enabling control of nonlinear systems. The objective is to extend the Linear Quadratic Regulator (LQR) method within the framework of interval analysis and develop a method for optimal control of systems with interval uncertainties.

In this method, the Pontryagin principle is initially applied to solve the interval quadratic systems and derive the necessary conditions for optimality. Subsequently, algebraic methods are transformed into ODEs. Finally, these equations are solved using the interval Chebyshev method, resulting in a confidence interval for controlling the closed-loop system.

By utilizing the Pontryagin principle and transforming the algebraic methods to ODEs, this approach provides a systematic way to tackle OCPs with interval uncertainties. The interval Chebyshev method further enhances the solution process by providing a reliable confidence interval for closed-loop control.

Consider a linear multivariable state-space model of the plant dynamics with interval uncertainties as follows:

$$\dot{x}(t) = \tilde{A}x(t) + \bar{B}u(t), \tag{6.91}$$

where $x(t) \in \mathbb{R}^n$ is a state vector and $u(t) \in \mathbb{R}^p$ an input vector. The elements $a_{i,j}, b_{i,j}$ ($i, j = 1, 2, ..., n; k = 1, 2, ..., p$) of matrix $\tilde{A} \in I(\mathbb{R}^{n \times n})$ and matrix $\tilde{B} \in I(\mathbb{R}^{n \times p})$ are interval integers bounded in a defined upper and lower interval, i.e., $\tilde{A} = [\underline{A}, \overline{A}]$ and $\tilde{B} = [\underline{B}, \overline{B}]$ are interval system matrix and input matrix, where their elements have lied between upper and lower bounds.

The boundary conditions of the system are

$$x(t_0) = X_0, x(t_f) = X_f. \tag{6.92}$$

where X_0 and X_f describe the initial and the final states of the system, respectively. Consider the performance measure as follows:

$$J(x(t), u(t), \Delta) = \frac{1}{2} x^T(t_f) x(t_f) + \frac{1}{2} \int_{t_0}^{t_f} [x^T(t) u^T(t)] \begin{bmatrix} \tilde{Q}(t) & 0 \\ 0 & \tilde{R}(t) \end{bmatrix} \begin{bmatrix} x(t) \\ u(t) \end{bmatrix} dt, \tag{6.93}$$

where Δ signifies the system interval uncertainties, $\tilde{Q}(t)$ describes the positive semi-definite, $\tilde{R}(t)$ specifies positive definite interval matrices, and $J(t, x(t), u(t), \Delta) = [\underline{j}(t, x(t), u(t)), \bar{j}(t, x(t), u(t))]$ describes the interval-valued performance index.

By expanding the interval arithmetic into the Pontryagin principle, the interval Hamiltonian equation of the problem is achieved as follows:

$$\tilde{H}(x(t), u(t), \lambda(t)) = \frac{1}{2} x^T(t) \tilde{Q} x(t) R U(t) + \lambda(t) (\tilde{A} x(t) + \tilde{B} u(t)), \tag{6.94}$$

By applying the optimal control to the interval Hamiltonian matrix,

$$\frac{\partial \tilde{H}}{\partial u} = \tilde{0} \rightarrow \tilde{R} u(t) + \tilde{B}^T \lambda = 0, \tag{6.95}$$

$$\Rightarrow \tilde{u}^*(t) = -\tilde{R}^{-1} \tilde{B}^T \lambda,$$

$$\dot{x}(t) = \frac{\partial \tilde{H}}{\partial \lambda} \rightarrow \dot{x}(t) = \tilde{A} x(t) + \tilde{B} u(t)$$

$$\dot{\lambda}(t) = -\frac{\partial \tilde{H}}{\partial x} \rightarrow ?\dot{\lambda}(t) = -\tilde{Q} x(t) - \tilde{A}^T \lambda(t) \tag{6.96}$$

where

$$\begin{bmatrix} \dot{x}(t) \\ \dot{\lambda}(t) \end{bmatrix} = \begin{bmatrix} \tilde{A} & -\tilde{E} \\ -\tilde{Q} & -\tilde{A}^T \end{bmatrix} \begin{bmatrix} x(t) \\ \lambda(t) \end{bmatrix},$$

$$\tilde{E} = \tilde{B}\tilde{R}^{-1}\tilde{B}^T,$$

(6.97)

Note that if $\tilde{R}$ is a matrix (i.e., multi-input systems), the inverse matrix of the equation should be first achieved by the explained interval inverse matrix method.

For closed-loop optimal control, we assume $\lambda(t) = \tilde{P}(t)x(t)$. Since,

$$\tilde{u}^*(t) = -\tilde{R}^{-1}\tilde{B}^T\tilde{P}x(t) = -\tilde{k}x(t)$$

(6.98)

and

$$\begin{cases} \dot{x}(t) = \tilde{A}x(t) - \tilde{B}\tilde{R}^{-1}\tilde{B}^T\tilde{P}x(t) \\ \dot{\lambda}(t) = -\tilde{Q}x - \tilde{A}^T\tilde{P}x(t) \end{cases}$$

(6.99)

By solving the equation above, the final equation will become as follows:

$$\tilde{P}\tilde{A} + \tilde{A}^T\tilde{P} + \tilde{P}\tilde{B}\tilde{R}^{-1}\tilde{B}^T\tilde{P}x + Q + \dot{\tilde{P}} = 0$$

(6.100)

This equation is the interval matrix extension of the Algebraic Riccati Equation (*IMARE*), where the interval solution $\tilde{P}$ is required to achieve the optimal interval feedback gain $\tilde{k}$ such that $\tilde{k} = \tilde{R}^{-1}\tilde{B}^T\tilde{P}$.

By simplification of the achieved IMARE, some IODEs have been extracted.

An interval method is required to solve the IODEs in this study. However, interval methods often suffer from overestimation in computation due to the inherent wrapping effect [26]. To address this issue, researchers have been working on methods to reduce the overestimation of interval calculations [27, 28].

In this study, the Chebyshev inclusion function is employed due to its capability to tighten the interval bounds [29]. The Chebyshev inclusion function has been discussed in detail in the preceding section. By using this function, the final solutions provide an interval value for the feedback coefficients. These coefficients represent a confident bound and enable the derivation of a suboptimal controller under interval uncertainties.

Once the feedback gain coefficients are obtained, the final feedback law can be derived using Eq. (6.35). This completes the process of obtaining the feedback control law based on the interval optimization approach, where the Chebyshev inclusion function plays a crucial role in tightening the interval bounds and reducing the overestimation in the computation.

Case Study 6.8 Optimal Control of a DC Motor with Interval Uncertainties

DC motors have long been widely utilized as primary actuators in various industrial applications due to their simple characteristics and stability. The speed of DC motors is typically proportional to the applied voltage, and there are several techniques available for controlling their speed, including electronic controllers and battery trapping methods (as mentioned in [30, 31]).

However, a significant problem often overlooked in the mathematical modeling of DC motors is the neglect of unknown factors, such as changes in resistance values caused by temperature fluctuations. This omission can result in the design of a controller that does not fully account for the system, posing challenges in ensuring robust performance. Therefore, incorporating interval analysis can enhance the robustness and practicality of the controller against variations in the motor's conditions. Figure 6.12 shows the schematic representation of a DC motor with interval parameters (~):

By considering the interval uncertainty in the modeling of the DC motor, we can capture the potential variations in resistance values and other parameters, allowing for a more comprehensive and robust design of the controller. This approach improves the controller's ability to handle uncertainties and changes in the motor's operating conditions, ultimately enhancing the performance and reliability of the overall system. Consider the motor toque (T_m). This term is related to the armature current (i_a) with the following formula:

$$\tilde{T}_m = \tilde{K}_i \tilde{i}_a, \tag{6.101}$$

The back emf ($\tilde{e}_b$) is also relative to angular velocity ($\tilde{\omega}_m$) by

$$\tilde{e}_b = \tilde{K}_b \omega_m = \tilde{K}_b \frac{d\theta}{dt}, \tag{6.102}$$

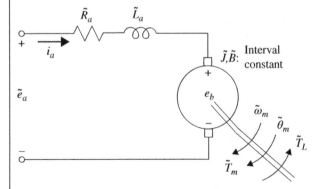

Figure 6.12 Schematic of DC motor with interval parameters.

By applying Newton's law and Kirchoff's law, the system equations can be achieved as follows:

$$\tilde{L}_a \frac{d\tilde{i}_a}{dt} + \tilde{R}_a \tilde{i}_a = \tilde{e}_a - \tilde{K}_b \frac{d\theta}{dt},$$

$$\tilde{J}_m \frac{d^2\theta}{dt^2} + \tilde{B}_m \frac{d\theta}{dt} = \tilde{K}_i \tilde{i}_a. \tag{6.103}$$

$$\tilde{e}_b = \tilde{K}_b \omega_m = \tilde{K}_b \frac{d\theta}{dt}, \tag{6.104}$$

From the equations above, the state space of the system can be considered as follows:

$$\begin{bmatrix} \dot{\tilde{i}}_a \\ \dot{\tilde{\omega}}_m \\ \dot{\tilde{\theta}}_m \end{bmatrix} = \begin{bmatrix} -\tilde{R}_a/\tilde{L}_a & -\tilde{K}_b/\tilde{L}_a & \{0\} \\ \tilde{K}_i/\tilde{J}_m & -\tilde{B}_m/\tilde{J}_m & \{0\} \\ \{0\} & \{1\} & \{0\} \end{bmatrix} \begin{bmatrix} \tilde{i}_a \\ \tilde{\omega}_m \\ \tilde{\theta}_m \end{bmatrix} + \begin{bmatrix} 1/\tilde{L}_a \\ \{0\} \\ \{0\} \end{bmatrix} \tilde{e}_a, \tag{6.105}$$

$$\tilde{\omega}_m = [\{0\} \ \{1\} \ \{0\}] \begin{bmatrix} \tilde{i}_a \\ \tilde{\omega}_m \\ \tilde{\theta}_m \end{bmatrix}. \tag{6.106}$$

where the terms $\{0\} = [0, 0]$ and $\{1\} = [1, 1]$ called, in turn, degenerate zero and degenerate one integer [32]. Table 6.4 illustrates the DC motor parameters with interval values.

Table 6.4 DC motor parameters with interval values.

Symbol	Value with interval uncertainty
$\tilde{E}$	$[11, 13]$ (volt)
$\tilde{J}_m$	$[0.008, 0.02]$ (kgm^2)
$\tilde{B}_m$	$[0.00002, 0.00005]$ (kgm^2/s)
$\tilde{K}_i$	$[0.021, 0.024]$ (Nm/A)
$\tilde{K}_b$	$[0.020, 0.024]$ (V/rad/s)
$\tilde{R}_a$	$[0.5, 1.5]$ (Ω)
$\tilde{L}_a$	$[0.25, 0.75]$ (H)

Consider the OCP for the DC motor speed model with interval uncertainties, as described in the previous section. The model can be represented by a linear differential equation with $n = 3$, $p = 2$, and interval matrices $\tilde{A}$ and $\tilde{B}$. Before verifying the validity of the LQR controller, it is crucial to analyze the system's controllability. However, controllability in interval matrices is different from that in real matrices. Therefore, an interval-based method is introduced and utilized to analyze the controllability of the DC motor in the following section.

- *Controllability test for the proposed DC motor with interval uncertainties*

In the subsequent equations, we will apply the designed controller to the nonlinear system, which is represented by an interval model. Therefore, it is essential to examine the controllability of the nonlinear model as well.

Consider Eq. (6.28). The system is controllable if for the initial condition of x $(0) = x_0$ and for any given vector x_f, there exists a limited time like t_f and input u (t) in the interval $[0, t_f]$ where this input mapping the system from the x_0 into x_f in time t_f; i.e., $x(t_f) = x_f$. Otherwise, the given equation is uncontrollable. There are different methods for analyzing the controllability of the systems with real integer values [33]. Consider the introduced third-order interval DC with the following interval parameters:

$$\tilde{A} = \begin{bmatrix} [-6, -0.6] & [-0.096, -0.027] & \{0\} \\ [1.05, 3] & [-0.0062, -0.001] & \{0\} \\ \{0\} & \{1\} & [\{0\}] \end{bmatrix}$$

$$\tilde{B} = \begin{bmatrix} [4/3, 4] \\ \{0\} \\ \{0\} \end{bmatrix}, \tilde{C} = [\{0\} \quad \{1\} \quad \{0\}]. \tag{6.107}$$

The system $\left(\tilde{A}, \tilde{B}\right)$ will be controllable if the controllability matrix $\tilde{R} = \left[\tilde{B}\tilde{A}\tilde{B}\tilde{A}^2\tilde{B}\tilde{A}^{n-1}\tilde{B}\right]$ has full row rank (i.e., $\rho(\tilde{R}) = \{n\} = [n, n]$). By extending the interval arithmetic to this assumption and calculating the system controllability matrix, we have

$$\rho(\tilde{R}) = \rho \left(\begin{bmatrix} 1.33,4 & [-24,6.4] & [-86.74,114.15] \\ \{0\} & [-1.2,12] & [-72.07,36.39] \\ \{0\} & \{0\} & [-1.20,12] \end{bmatrix} \right) = \{3\} \tag{6.108}$$

From the above, it is clear that the studied system is controllable, and hence its closed-loop poles can be placed anywhere in the s-plane.

Optimal Control of the Interval Valued DC Motor Using Proposed Interval LQR

To design the optimal control of the considered interval LQR, we should first specify the augmented system matrices. From the previous subsection, we have $\tilde{A}$ and $\tilde{B}$ matrices. The other required matrices are given in the following:

$$\tilde{Q}(t) = \begin{bmatrix} \tilde{\gamma}_1 & \tilde{\gamma}_3 & \tilde{\gamma}_3 \\ \tilde{\gamma}_3 & \tilde{\gamma}_1 & \tilde{\gamma}_3 \\ \tilde{\gamma}_3 & \tilde{\gamma}_3 & \tilde{\gamma}_2 \end{bmatrix}, \tilde{R} = \{8\}$$

$$t_0 = 0, t_f = 10, F(t_f) = 0$$

(6.109)

where $\tilde{\gamma}_1 = [1/2, 2], \tilde{\gamma}_2 = [1, 2]$, and $\tilde{\gamma}_3 = \{1/2\}$.

By considering the given matrices and after applying them to the interval Algebraic Riccati Equation, the following interval differential system is achieved:

$$\dot{\tilde{P}}_1 = \tilde{\alpha}_1 \tilde{P}_1 + \tilde{\alpha}_2 \tilde{P}_2 + \tilde{\beta} \tilde{P}_1^2 - \tilde{\gamma}_1$$

$$\dot{\tilde{P}}_2 = \tilde{\alpha}_3 \tilde{P}_2 + \tilde{\alpha}_4 \tilde{P}_1 + \tilde{\alpha}_5 \tilde{P}_4 - \tilde{P}_3 + \tilde{\beta} \tilde{P}_1 \tilde{P}_2 - \tilde{\gamma}_3$$

$$\dot{\tilde{P}}_3 = \tilde{\alpha}_6 \tilde{P}_3 + \tilde{\alpha}_7 \tilde{P}_5 + \tilde{\beta} \tilde{P}_1 \tilde{P}_3 + \tilde{\gamma}_3$$

$$\dot{\tilde{P}}_4 = 2\tilde{\alpha}_4 \tilde{P}_2 + \tilde{\alpha}_8 \tilde{P}_4 - 2\tilde{P}_5 + \tilde{\beta} \tilde{P}_2^2 - \tilde{\gamma}_1$$

$$\dot{\tilde{P}}_5 = \tilde{\alpha}_4 \tilde{P}_3 + \tilde{\alpha}_8 \tilde{P}_5 - \tilde{P}_6 + \tilde{\beta} \tilde{P}_2 \tilde{P}_3 - \tilde{\gamma}_3$$

$$\dot{\tilde{P}}_6 = \tilde{\beta} \tilde{P}_3^2 - \tilde{\gamma}_2$$

(6.110)

where

$$\tilde{\alpha}_1 = [1.2, 12], \tilde{\alpha}_2 = [-6, -2.1], \tilde{\alpha}_3 = [0.601, 6.0062],$$

$$\tilde{\alpha}_4 = [0.027, 0.096] \tilde{\alpha}_5 = [-3, -1.05] \tilde{\alpha}_6 = [0.6, 6]$$

$$\tilde{\alpha}_7 = [0.002, 0.0124], \tilde{\alpha}_8 = [0.001, 0.0062], \tilde{\beta} = [1/9, 1].$$

By applying the Chebyshev inclusion method to the ODE system, the optimum feedback gain $(K = [\underline{k}, \overline{k}])$ is obtained as follows: $\underline{k} = [0.3638, 0.8451, 0.4319]$, $\overline{k} = [0.9009, 0.8384, 0.4585]$. The eigenvalues for different closed-loop interval controllers are given in Table 6.5.

Numerous methods have been proposed for obtaining a robust optimal control, as discussed in [34, 35]. However, these methods often suffer from a significant drawback in that they design the final controller solely based on real-valued arithmetic centered around the interval, neglecting the associated confidence interval. In contrast, our approach aims to consider the entire

Table 6.5 Eigenvalues for different closed-loop interval controllers.

	Eigenvalues		
Plant	λ_1	λ_2	λ_3
Lower bounds	$-5.0518 + 0.0000i$	$-0.6030 + 0.5449i$	$-0.6030 - 0.5449i$
Intermediate	$-4.0342 + 0.0000i$	$-0.3432 + 0.3406i$	$-0.3432 - 0.3406i$
Intermediate	$-2.6723 + 0.0000i$	$-0.7530 + 0.6674i$	$-0.7530 - 0.6674i$
Upper bounds	$-1.4909 + 0.0000i$	$-0.6271 + 0.7666i$	$-0.6271 - 0.7666i$

confidence interval comprehensively, thereby mitigating potential sudden issues in the system. To demonstrate the effectiveness of our method, we compare it with the Monte Carlo (MC) method, using 50 iterations, based on the proposed Iterative Linear Quadratic Regulator (ILQR) method. The comparison results are depicted in Figure 6.13.

This figure illustrates the performance and reliability achieved with our ILQR method compared to the MC method. By considering the entire confidence interval, our approach provides a more robust and stable control solution, minimizing the risk of unexpected problems in the system. This highlights the advantage of incorporating interval analysis in the design process, ensuring a controller that can confidently handle uncertainties and deliver consistent performance.

It is observed that the proposed MC interval method incorporates the centered method. Furthermore, the Chebyshev inclusion method is utilized as the primary approach to reduce the overestimation of the interval and determine the optimal interval gain for closed-loop control of the DC motor under interval uncertainties.

Figure 6.14 showcases the step response of the DC motor, providing a visual representation of the performance achieved. In this study, we compare the proposed Chebyshev inclusion method with the MC method using 1000 iterations. Analyzing the results reveals that the interval obtained from the Chebyshev inclusion method is narrower compared to that obtained from the MC method. This indicates superior control performance with the Chebyshev inclusion method.

The narrower interval in the Chebyshev inclusion method implies a more precise and confident representation of the system's behavior. This improved

Figure 6.13 Comparison of the proposed MC method (50 iterations) with lower and upper bounds and the centered method: (a) step response for the DC motor with interval uncertainties. (b) Focus of the (a) in the time interval [0, 12].

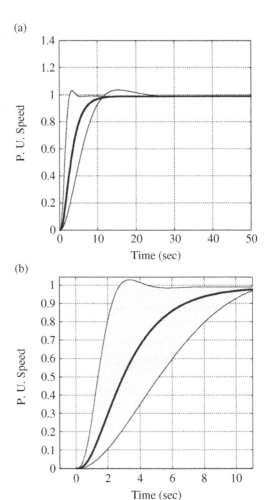

precision translates into better control performance, allowing for more accurate tracking of desired trajectories and enhanced stability. By effectively reducing the overestimation of intervals, the proposed Chebyshev inclusion method demonstrates its effectiveness in addressing the challenges posed by interval uncertainties in controlling DC motors.

The obtained results clearly illustrate the efficacy of the proposed Chebyshev inclusion method in achieving a tighter interval compared to the MC simulation method. This improved performance can be attributed to the proposed

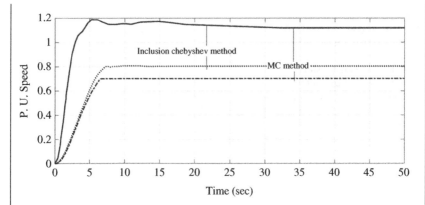

Figure 6.14 Step response for DC motor with uncertainties by MC and Chebyshev inclusion methods.

method's ability to effectively manage the overestimation issue caused by the interval optimal control's wrapping effect. Furthermore, the proposed Chebyshev inclusion method demonstrates reasonable computational efficiency, offering a lower computational cost compared to the widely used MC method. To provide quantitative insights, Table 6.6 presents the calculation times for both the Chebyshev inclusion-based and MC-based interval methods. The table indicates that the Chebyshev inclusion method requires less running time than the MC method. These findings highlight the advantages of the Chebyshev inclusion method not only in terms of achieving more accurate modeling of interval uncertainties but also in terms of computational efficiency. By effectively addressing the overestimation issue and offering reduced computational time, the proposed method presents itself as a promising approach for robust control of DC motors under interval uncertainties.

The state and control variables of the Chebyshev inclusion method for the system are shown in Figure 6.15.

Table 6.6 Running time of the proposed method.

Method	Chebyshev inclusion method	MC method
Running time (sec.)	10.2	1000

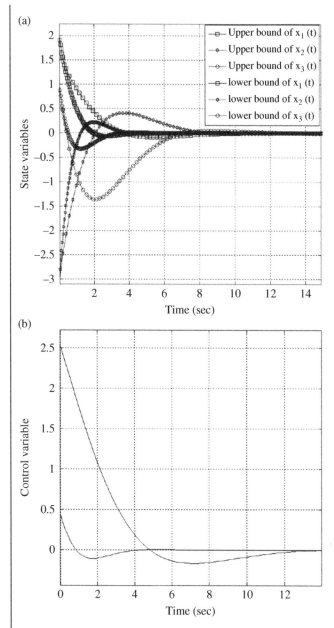

Figure 6.15 Optimal state (a) and control (b) variables for DC motor with interval uncertainties.

Case Study 6.9   Approximated Two-Wheeled and Self-Balancing Robot Mathematical Model with Interval Uncertainties

---

The self-balancing robot investigated in this research comprises a chassis mounted on top of an axle with two wheels, without any additional balancing support. Due to its resemblance to a vehicle-mounted inverted pendulum, the dynamic system analysis process for this robot is more intricate. One crucial aspect often overlooked in the mathematical modeling of two-wheeled self-balancing robots is the omission of unknown terms, such as resistance values. Neglecting these unknowns can result in discrepancies in the design. To address this issue, a novel method based on interval analysis is employed in this study to account for uncertainties in robot identification. This method enhances the robustness and practicality of the controller in the face of changing conditions. The modeling process in the paper begins with the separation of wheel and pendulum analyses [36], allowing for a more comprehensive understanding of the system's dynamics. Subsequently, potential uncertainties are incorporated into the system model.

By integrating interval analysis and accounting for uncertainties, the proposed method enables a more accurate representation of the self-balancing robot's behavior. It empowers the controller to handle unforeseen variations and enhances the overall stability and performance of the system. This approach represents a significant step toward developing robust and practical control strategies for self-balancing robots in real-world scenarios.

Figure 6.16 shows a diagram of revolver force analysis. According to the revolver, the force equation can be obtained according to Newton's law [37] and the rotational torque formula [38]:

$$M_w \ddot{x} = H_{fR} - H_R,$$
$$I_w \ddot{\theta}_w = C_R - H_{fR}.R. \tag{6.111}$$

where M_w represents the weight of the wheel, I_w is the moment of inertia of the wheel, R represents the radius of the wheel, and $\ddot{x}$ is the wheel acceleration of the x-axis.

The C_R is the right wheel torque and the H_R is the Z-axis force of the right wheel with the car body. The H_{fR} is the interatomic force of the right wheel with the ground and the term θ_w is the angle of the wheel around the Z axis direction.

Similarly, for the right wheel, the force equation can be achieved by the following formula:

$$M_w \ddot{x} = H_{LR} - H_R,$$
$$I_w \ddot{\theta}_w = C_L - H_{fL}.R. \tag{6.112}$$

Figure 6.16 The force analysis of revolver. *Source:* Fang et al. [36]/Jian Fang/ CC BY.

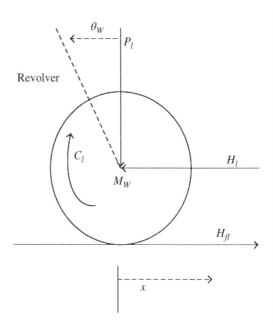

where the C_L is the left wheel torque and H_L is the Z-axis force of the left wheel with the car body.H_{fL} is the interatomic force of the left wheel with the ground. The final equation will be achieved as follows:

$$2\left(M_w + \frac{I_w}{R^2}\right)\ddot{x} = \frac{C_R + C_L}{R} - (H_R + H_L). \tag{6.113}$$

Figure 6.16 displays the force analysis of the revolver [36].

After using Newton's second law for the horizontal and vertical forces [36], the final approximated interval system is achieved as follows:

$$\left(M_p + 2M_W \frac{I_w}{R^2}\right)\ddot{x} = \frac{C_R + C_L}{R} - 2M_p l\theta_P,$$
$$\left(2M_p l^2 + I_p\right)\ddot{\theta}_P = M_p g l\theta_P - M_p l \ldots x. \tag{6.114}$$

where θ_P represents the angle of the car body deviating from the Z axis direction, M_P is the weight of the car body, I_P is the moment of inertia of the car body, and l is the height at which the car body is apart from the shaft.

In this equation, the output torque for the wheel is $C_R = C_L = I_R(dw/dt) = (k_m/R)U_a - (-k_m k_e/R)\dot{\theta}_w$; finally, the state space representation of the two-wheeled self-balancing robot is achieved as follows:

$$
\begin{bmatrix} \dot{x} \\ \ddot{x} \\ \dot{\theta}_p \\ \ddot{\theta}_p \end{bmatrix} = \begin{bmatrix} \{0\} & \{1\} & \{0\} & \{0\} \\ \{0\} & \dfrac{2k_m k_e(M_P lr - I_P - M_P l^2)}{Rr^2 A} & \dfrac{M_P^2 g l^2}{A} & \{0\} \\ \{0\} & \{0\} & \{0\} & \{1\} \\ \{0\} & \dfrac{2k_m k_e(rB - M_P l)}{Rr^2 A} & \dfrac{M_P^2 g l B}{A} & \{0\} \end{bmatrix} \begin{bmatrix} x \\ \dot{x} \\ \theta_p \\ \dot{\theta}_p \end{bmatrix}
$$

$$
+ \begin{bmatrix} \{0\} \\ \dfrac{2k_m(I_P + M_P l^2 - M_P lr)}{RrA} \\ \{0\} \\ \dfrac{2k_m(M_P l - rB)}{RrA} \end{bmatrix} U_a.
$$

(6.115)

And the output state representation is

$$
y = [\{0\}\{0\}\{1\}\{0\}] \begin{bmatrix} x \\ \dot{x} \\ \theta_p \\ \dot{\theta}_p \end{bmatrix}.
$$

(6.116)

where $A = [I_P \beta + 2M_P l^2(M_w + (I_w/r^2))]$ and $B = [2M_w + (2I_w/r^2) + M_P$ and the terms $\{0\} = [0, 0]$ and $\{1\} = [1, 1]$ called degenerate integers [32].

The final interval state representation of the system with uncertainties is given below:

$$
\tilde{A} = \begin{bmatrix} \{0\} & \{1\} & \{0\} & \{0\} \\ \{0\} & [-0.25, -0.11] & [24.6, 56.05] & \{0\} \\ \{0\} & \{0\} & \{0\} & \{1\} \\ \{0\} & [-0.61, -0.49] & [237, 239] & \{0\} \end{bmatrix}
$$

(6.117)

$$
B = \begin{bmatrix} \{0\} \\ [0.41, 0.56] \\ \{0\} \\ [1.9, 2.72] \end{bmatrix}, \quad C = [\{0\} \quad \{0\} \quad \{1\} \quad \{0\}]
$$

6.11   Problem Statement (Interval Quadratic Regulator)

Consider a linear multivariable state-space model of the plant dynamics with interval uncertainties as follows:

$$\dot{x}(t) = \tilde{A}x(t) + \tilde{B}u(t), \tag{6.118}$$

where $x(t) \in \mathbb{R}^n$ is a state vector and $u(t) \in \mathbb{R}^p$ is an input vector. The elements $a_{i,j}$, $b_{i,j}(i, j = 1, 2, ..., n; k = 1, 2, ..., p)$ of matrix $\tilde{A} \in I(\mathbb{R}^{n \times n})$ and matrix $\tilde{B} \in I(\mathbb{R}^{n \times p})$ are interval integers bounded in a defined upper and lower interval, i.e., $\tilde{A} = [\underline{A}, \overline{A}]$ and $\tilde{B} = [\underline{B}, \overline{B}]$ are interval system matrix and input matrix where their elements have lied between upper and lower bounds.

The boundary conditions of the system are

$$x(t_0) = X_0, x(t_f) = X_f. \tag{6.119}$$

where X_0 and X_f describe the initial and the final states of the system, respectively.

Consider the performance measure as follows:

$$
\begin{aligned}
J(x(t), u(t), \Delta) &= \frac{1}{2}x^T(t_f)F(t_f)x(t_f) \\
&+ \frac{1}{2}\int_{t_0}^{t_f} \left[x^T(t) u^T(t)\right] \begin{bmatrix} \tilde{Q}(t) & 0 \\ 0 & \tilde{R}(t) \end{bmatrix} \begin{bmatrix} x(t) \\ u(t) \end{bmatrix} dt
\end{aligned} \tag{6.120}
$$

In this case, Δ signifies the system uncertainties, $\tilde{Q}(t)$ is positive semi-definite, and $\tilde{R}(t)$ is positive definite interval matrices and $J(t, x(t), u(t), \Delta) = \left[\underline{j}(t, x(t), u(t)), \overline{j}(t, x(t), u(t))\right]$ describes the interval-valued performance index.

By expanding the interval arithmetic into the Pontryagin principle, the interval Hamiltonian equation of the problem is achieved as follows:

$$\tilde{H}(x(t), u(t), \lambda(t)) = \frac{1}{2}x^T(t)\tilde{Q}x(t) + \frac{1}{2}u^T(t)\tilde{R}u(t) + \lambda(t)\left(\tilde{A}x(t) + \tilde{B}u(t)\right), \tag{6.121}$$

By applying the optimal control on the interval Hamiltonian matrix,

$$
\begin{aligned}
\frac{\partial \tilde{H}}{\partial u} &= \tilde{0} \rightarrow \tilde{R}u(t) + \tilde{B}^T \lambda = 0, \\
&\Rightarrow \tilde{u}^*(t) = -\tilde{R}^{-1}\tilde{B}^T \lambda,
\end{aligned} \tag{6.122}
$$

$$
\begin{aligned}
\dot{x}(t) &= \frac{\partial \tilde{H}}{\partial \lambda} \rightarrow \dot{x}(t) = \tilde{A}x(t) + \tilde{B}u(t) \\
\dot{\lambda}(t) &= -\frac{\partial \tilde{H}}{\partial x} \rightarrow \dot{\lambda}(t) = -\tilde{Q}x(t) - \tilde{A}^T \lambda(t)
\end{aligned} \tag{6.123}
$$

That is,

$$
\begin{bmatrix} \dot{x}(t) \\ \dot{\lambda}(t) \end{bmatrix} = \begin{bmatrix} \tilde{A} & -\tilde{E} \\ -\tilde{Q} & -\tilde{A}^T \end{bmatrix} \begin{bmatrix} x(t) \\ \lambda(t) \end{bmatrix},
$$
$$
\tilde{E} = \tilde{B}\tilde{R}^{-1}\tilde{B}^T,
$$

(6.124)

Note that if $\tilde{R}$ is a matrix (i.e., multi-input systems), the inverse matrix of the equation should be first achieved by the explained interval inverse matrix method. For the closed-loop optimal control, we assume $\lambda(t) = \tilde{P}(t)x(t)$. Since,

$$
\tilde{u}^*(t) = -\tilde{R}^{-1}\tilde{B}^T\tilde{P}x(t) = -\tilde{k}x(t)
$$

(6.125)

and

$$
\begin{cases} \dot{x}(t) = \tilde{A}x(t) - \tilde{B}\tilde{R}^{-1}\tilde{B}^T\tilde{P}x(t) \\ \dot{\lambda}(t) = -\tilde{Q}x - \tilde{A}^T\tilde{P}x(t) \end{cases}
$$

(6.126)

By solving the equation above, the final equation will be as follows:

$$
\tilde{P}\tilde{A} + \tilde{A}^T\tilde{P} + \tilde{P}\tilde{B}\tilde{R}^{-1}\tilde{B}^T\tilde{P}x + Q + \dot{\tilde{P}} = 0
$$

(6.127)

This equation is the interval matrix extension of the Algebraic Riccati Equation (*IMARE*), where the interval solution $\tilde{P}$ is required to achieve the optimal interval feedback gain $\tilde{k}$ such that $\tilde{k} = \tilde{R}^{-1}\tilde{B}^T\tilde{P}$.

By simplification of the achieved IMARE, some IODEs have been extracted.

To solve the IODEs, an interval method is necessary. However, interval methods often suffer from overestimation during computation due to their inherent wrapping effect [26]. Recognizing this drawback, researchers have made efforts to reduce overestimation in interval calculations [27, 28]. In this study, the Chebyshev inclusion function is employed due to its effectiveness in tightening the interval bounds [29]. Further details regarding the Chebyshev inclusion function can be found in the previous section. As a result, the solutions obtained in this study provide an interval value for feedback coefficients. This interval represents a confident bound that yields a suboptimal controller under interval uncertainties.

By utilizing the Chebyshev inclusion function and considering the resulting interval bounds, this study contributes to the development of robust control strategies that can effectively handle interval uncertainties. The provided feedback law ensures stability and performance in the presence of these uncertainties, offering practical solutions for real-world applications.

Case Study 6.10   Applying the Proposed Method on the Two-Wheeled and Self-Balancing Robot with Interval Uncertainties

---

Consider an OCP for the two-wheeled and self-balancing robot motor speed model with interval uncertainties, which is described in the previous section. A linear differential equation with $n = 4$, $p = 1$ and interval matrices $\tilde{A}$ and $\tilde{B}$. Despite the quadratic systems have controllability guarantees, because of the system uncertainties, we need to check its constructability based on interval methods like [39].

Controllability Testing of the Two-Wheeled, Self-Balancing Robot with Interval Uncertainties

Consider Eq. (6.28). This system will be controllable if for the initial condition of $x(0) = x_0$ and for any given vector x_f, there exists a limited time like t_f and input $u(t)$ in the interval$[0, t_f]$ where this input mapping the system from thex_0 into x_f in time t_f; i.e., $x(t_f) = x_f$; otherwise, the given equation is uncontrollable. There are different methods for analyzing the controllability of the systems with real integer values [33]. In this section, we utilized an interval-based method from Shashikhin, which is explained in the previous sections [39]. From the above, for the case study, we have

$$\tilde{D}= \left[\tilde{B}\,|\,\tilde{A}\tilde{B}\,|\,\tilde{A}^2\tilde{B}\,|\,\tilde{A}^3\tilde{B}\right]$$

$$= \begin{bmatrix} \{0\} & [0.41,0.56] & [-0.14,-0.045] & [46.74,152.49] \\ [0.41,0.56] & [-0.14,-0.0451] & [46.74,152.46] & [-57.27,-10.09] \\ \{0\} & [1.9,2.72] & [-0.34,0.20] & [450.32,650.16] \\ [1.9,2.72] & [-0.34,-0.2] & [450.32,650.16] & [-174.66,-70.51] \end{bmatrix}.$$

$$(6.128)$$

$$\tilde{D}\tilde{D}^T = \begin{bmatrix} [2185,23254] & [-8754.5,-4737] & [21049,91440] & [-26725,-3316] \\ [-8754.5,-473.7] & [2287,26529] & [-37287,-4553] & [21760,109150] \\ [21049,99144] & [-37287,-4553] & [202790,422720] & [-113780,-31840] \\ [-26640,-3316] & [21760,108830] & [-113780,-31840] & [207760,453220] \end{bmatrix}.$$

$$(6.129)$$

By achieving the median value,

$$med\left(\tilde{D}\tilde{D}^{T}\right) = \begin{bmatrix} 12720 & -6746 & 56245 & -15021 \\ -4614 & 14408 & -20920 & 65455 \\ 60097 & -20920 & 312755 & -72810 \\ -14978 & 65295 & -72810 & 330490 \end{bmatrix}. \quad (6.130)$$

After applying the formula in the flowchart, the eigenvalues and the null values are achieved as follows:

$$\lambda_1 = 411360; \varepsilon_1 = 3.6007e^{-10} \quad \lambda_3 = 3170; \varepsilon_3 = 1.6130e^{-11}$$
$$\lambda_2 = 255930; \varepsilon_2 = 2.8086e^{-10} \quad \lambda_4 = 470; \varepsilon_4 = 4.7749e^{-11}$$

And the final interval eigenvalues are $\tilde{\lambda}_1 = [411360, 411360], \tilde{\lambda}_2 = [3170, 3170], \tilde{\lambda}_3 = [255930, 255930]$, and $\tilde{\lambda}_4 = [470, 470]$. As can be seen, the entire interval eigenvalues are positive so the system is controllable.

Note that because of the low amount of epsilon and the high amount of the eigenvalues, the lower and the upper bounds of the interval eigenvalues are similar to the degenerate integers.

Optimal Interval Control of the Two-Wheeled and Self-Balancing Robot

To design the optimal control of the considered interval LQR, we should first specify the augmented system matrices [40]. From the previous subsection, we have $\tilde{A}$ and $\tilde{B}$ matrices. The other required matrices are given in the following:

$$\tilde{Q}(t) = \begin{bmatrix} [220,223] & \{0\} & \{0\} & \{0\} \\ \{0\} & [168,172] & \{0\} & \{0\} \\ \{0\} & \{0\} & [120,122.2] & \{1\} \\ \{0\} & \{0\} & \{0\} & [187,188.6] \end{bmatrix}, \quad (6.131)$$

$$\tilde{R} = \{1.77\}.$$

By considering the given matrices and generating the Interval Algebraic Riccati Equation and solving these equations using the proposed interval Chebyshev method, the optimum feedback gain ($K = [\underline{k}, \overline{k}]$) is obtained as follows:

$$\underline{k} = [-11.19, -16.86, 221.39, 19.90], \quad \overline{k} = [-11.19, -15.20, 295.69, 23.55].$$

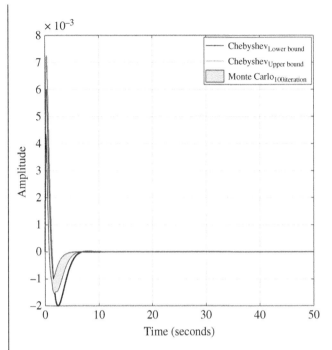

Figure 6.17 Comparison of step response for proposed interval Chebyshev method and MC method (100 iterations) for the two-wheeled and self-balancing robot with interval uncertainties.

Here, we consider all the confidence intervals for preventing the system from sudden problems. In the following, the simulation results of the proposed method and its comparison with the MC method are given. The iteration for the MC method is set to 100. Figure 6.17 illustrates the comparison of step response for the proposed Interval Chebyshev method and MC method (100 iterations) for the two-wheeled and self-balancing robot with interval uncertainties.

From the above results, it can be seen that while the upper bound of the proposed method and MC with 100 iterations is equal, the lower bound of the proposed method includes more surface than the MC. The reason is that the MC method depends on the number of its iterations.

In the following, we consider two random values for the system close to the lower bound and analyze the results. From the figure, it can be seen that the *Random₁* is included in the lower bound of both MC and the proposed method and the *Random₁* has a small error (about $1e-3$) in the lower bound for about 1.5 seconds and then it stands on the guaranteed interval (Figure 6.18).

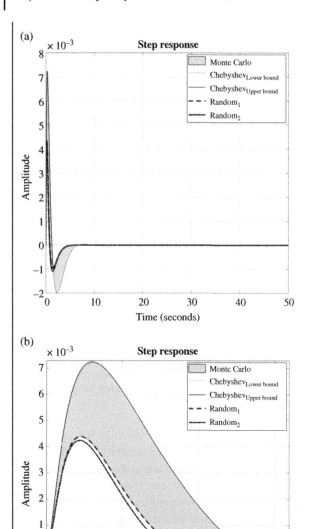

Figure 6.18 Step response for two-wheeled and self-balancing robot with uncertainties by MC and Chebyshev inclusion methods and two random inputs: (a) in the time interval [0, 50] and (b) [0, 1.5].

Table 6.7 Running time of the proposed method.

Method	Chebyshev inclusion method	MC method
Running time (sec.)	1.2	30

From the results, it can be concluded that the proposed Chebyshev inclusion method can achieve the guaranteed interval for the system. The proposed method has a reasonable computational cost that is less than the widely used MC method. The calculation time for the Chebyshev inclusion-based and MC-based interval methods is illustrated in Table 6.7.

It can be seen that the Chebyshev inclusion method requires less running time than the MC method.

6.12 Interval Optimal Control Based on Direct Method

- *Chebyshev orthogonal functions*

Based on the previous explanations, the indirect method enables the solving of a specific problem by defining a particular solution and utilizing separate mathematical methods for each OCP. On the other hand, the direct method provides a more general solution that can be applied to any type of optimized control problem, regardless of its complexity. In addition, in an indirect method, the OCP must be first converted into a unit cost function. That is, if we have several specific goals, we should reduce them into a single function. But in the direct method, the multi-objective optimization method can be used for any problem. Consider the unconstrained OCP below:

$$\min \; J(t, X(t), U(t), \Delta) = \int_{t_0}^{t_f} G(t, X(t), U(t), \Delta) \, dt \qquad (6.132)$$

The system dynamics and its initial and final conditions are as follows:

$$X(t) = f(t, X(t), U(t), \Delta), t \in \left(t_0, t_f\right)$$
$$X(t_0) = X_0, X\left(t_f\right) = X_f \qquad (6.133)$$

where $U(t)$ and $X(t)$, are the control variable and the state variable, respectively. $\Delta = [\delta_1, \delta_2, ..., \delta_n]$ defines the system uncertainties, X_0 and X_f are the initial and the final state variables.

For the simplicity of the work without loss of universality, we assume that our time interval is based on the time conversion as mentioned earlier, i.e.,

$$J(X) = \frac{t_f - t_0}{2} \int_{t_0}^{t_f} G\left(\frac{(t_f - t_0)}{2} t + \frac{(t_f + t_0)}{2}, X(t), \dot{X}(t), \Delta\right) dt \qquad (6.134)$$

where

$$U(t) = F\left(\frac{(t_f - t_0)}{2} t + \frac{(t_f + t_0)}{2}, X(t), \dot{X}(t), \Delta\right) \qquad (6.135)$$

And the initial conditions are as follows:

$$X(t_0) = X_{-1}, X(t_f) = X_1, \qquad (6.136)$$

By changing the time interval into the interval $[-1, 1]$ and taking into account the interval uncertain coefficients $\alpha \approx (a_0, a_1, ..., a_m)^T$, the state variable can be estimated by Chebyshev's series:

$$[X_m]([t]) \approx \frac{1}{2}[a_0] + \sum_{j=1}^{m} [f]\,([a_i]T_i), i = 0, 1, ..., m \qquad (6.137)$$

where $[a_i] = [\underline{a}_i, \overline{a}_i]$. Finally, based on the relationship between the state variable and the control variable, the control rule will be obtained.

$$a_i = \frac{2}{K} \sum_{j=1}^{m} f\left(\cos(\theta_j)\right) \cos(j\theta_i), j = 0, 1, ..., m \qquad (6.138)$$

$$\theta_j = \frac{2i - 1}{2K} \pi, i = 0, 1, ..., K \qquad (6.139)$$

where $K > m$.

In the direct solution for normal OCPs, these coefficients can be obtained in the following way:

$$[X_m]([t]) \approx \frac{1}{2}[a_0] + \sum_{j=1}^{m} [f]([a_j], t), j = 0, 1, ..., m \qquad (6.140)$$

When dealing with a system that exhibits interval uncertainty, we encounter the challenge of having two cost functions. In this situation, if we were to employ classical methods, we would need to transform these two functions into a single function using the Lagrange coefficients method. However, this approach proves to be highly inefficient. Alternatively, we can utilize multiobjective optimization methods instead of relying on classical techniques.

In single-objective optimization, the primary objective is to optimize (minimize or maximize) a single performance index that fully represents the quality of the system's response. However, in certain cases, relying solely on one index to guide the optimization process may not be practical or sufficient. This necessitates the consideration of multiple cost functions or performance indicators, simultaneously optimizing all of them. Multiobjective optimization is a highly effective and widely employed research technique in the field of optimization [41].

• *Improving the Chebyshev orthogonal functions*

The method described in the previous section is very good, but the confidence interval is very high. The reason for this can be searched for a large number of unknown coefficients. Therefore, if we can reduce the number of unknown coefficients, our data will be greatly reduced.

We first consider the following estimation for the path variable:

$$[X_1]([t]) \approx \sum_{i=0}^{2} [a_i]\, T_i(t) \tag{6.141}$$

Note that T_i here is an interval.

By considering the boundary conditions,

$$[a_0] = \frac{[X_1](1) + [X_1](-1)}{2} \ominus_{gh} [a_2] \tag{6.142}$$

$$[a_1] = [X_1](1) \ominus_{gh} \frac{[X_1](1) + [X_1](-1)}{2} \tag{6.143}$$

By replacing the equations,

$$[X_1]([t]) = [a_2]T_2(t) + \left([X_1](1) \ominus_{gh} \frac{[X_1](1) + [X_1](-1)}{2}\right) T_1(t)$$
$$+ \frac{[X_1](1) + [X_1](-1)}{2} \ominus_{gh} [a_2] \tag{6.144}$$

$$J(X(t), U(t), \Delta) = \int_{t_0}^{t_f} G\left(\frac{t_f - t_0}{2} t + \frac{t_f + t_0}{2}, X(t), \dot{X}(t), \Delta\right) dt \tag{6.145}$$

In the next step, the approximation accuracy has been improved:

$$[X_2]([t]) = [X_1]([t]) + \sum_{i=1}^{3} [a_i]\, T_i(t) \tag{6.146}$$

And using the boundary conditions $[a_2] = \{0\}$, $[a_1] = -[a_3]$ and consequently

$$[X_2]([t]) = [X_1]([t]) + [a_3]T_3(t) \ominus_{gh} [a_3]T_1(t) \tag{6.147}$$

Afterward, the performance index and the control variable can be obtained.

By following this methodology, the problem can be solved with a high degree of accuracy. The proposed method is a recursive technique that can be described as follows:

$$[X_{n+1}](t) = [X_n](t) + \sum_{i=n}^{n+2} [a_i] T_i(t) \tag{6.148}$$

Consequently, using the boundary conditions and the above equation,

$$
\begin{aligned}
[X_{n+1}](-1) &= [X_n](-1) \\
&\Rightarrow [a_{n+2}]T_{n+2}(-1) \\
&+ [a_{n+1}]T_{n+1}(-1) + [a_n]T_n(-1) \\
&= 0
\end{aligned} \tag{6.149}
$$

$$
\begin{aligned}
[X_{n+1}](-1) &= [X_n](1) \\
&\Rightarrow [a_{n+2}]T_{n+2}(1) + [a_{n+1}]T_{n+1}(1) \\
&+ [a_n]T_n(1) = 0
\end{aligned} \tag{6.150}
$$

For evaluating the uncertain coefficients,

$$[a_n] = \frac{T_{n+1}(-1)T_{n+2}(1) - T_{n+1}(1)T_{n+2}(-1)}{T_n(-1)T_{n+1}(1) - T_n(1)T_{n+1}(-1)} \tag{6.151}$$

$$[a_{n+1}] = \frac{T_n(-1)T_{n+2}(1) - T_n(1)T_{n+2}(-1)}{T_n(1)T_{n+1}(-1) - T_n(-1)T_{n+1}(1)} [a_{n+2}] \tag{6.152}$$

Consequently,

$$
\begin{aligned}
[X_{n+1}](t) = [X_n](t) &+ [a_{n+2}]T_{n+2}(t) \\
&+ \frac{T_n(-1)T_{n+2}(1) - T_n(1)T_{n+2}(-1)}{T_n(1)T_{n+1}(-1) - T_n(-1)T_{n+1}(1)} [a_{n+2}]T_{n+1}(t) \\
&+ \frac{T_{n+1}(-1)T_{n+2}(1) - T_{n+1}(1)T_{n+2}(-1)}{T_n(-1)T_{n+1}(1) - T_n(1)T_{n+1}(-1)} [a_{n+2}]T_n(t)
\end{aligned}
$$

$$\tag{6.153}$$

Note that the multiobjective optimization method is used to calculate uncertain (nondeterministic) coefficients at each stage. Finally, the value of the control variable and the performance index are obtained. The flowchart diagram of the proposed method is shown in Figure 6.19.

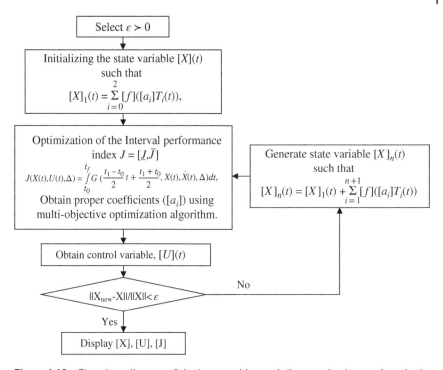

Figure 6.19 Flowchart diagram of the improved interval direct optimal control method.

6.13 Applied Simulations

Case Study 6.11 The Intercept Problem

The intercept problem holds great significance in the field of aeronautics, particularly in spacecraft applications. This issue arises when a follower spacecraft needs to intercept a target spacecraft. The objective of interception can vary depending on the situation, encompassing tasks such as rendezvous and docking, inspection, or even military interception.

To tackle this problem effectively, researchers have developed various techniques and algorithms that enable spacecraft to achieve successful interceptions. These methods involve intricate calculations and necessitate precise control over the follower spacecraft's trajectory and velocity. One notable technique is proportional navigation, which involves continuously adjusting the follower's velocity vector to align with the target. By doing so, the follower will eventually reach its target while optimizing fuel consumption.

Despite the advancements made in interception techniques, numerous challenges persist in intercepting spacecraft. These include uncertainties in orbital dynamics, communication delays between ground control and spacecraft, as well as limitations in propulsion systems.

To overcome these challenges and further enhance interception capabilities, ongoing research is actively exploring new techniques and technologies. These endeavors aim to improve the precision and reliability of interception maneuvers. More detailed information about these systems can be found in reference [42]. The accompanying figure illustrates two spacecraft, with the target labeled as A and the follower labeled as B.

As shown in Figure 6.20, two spacecraft are depicted – a target labeled A and a follower labeled B. The goal of B is to intercept A by following a specific trajectory while maintaining a safe distance from A until it reaches its destination. This figure illustrates how complex this process can be and highlights why it is such an important issue in aeronautics.

Overall, solving the intercept problem remains critical for space exploration missions as well as for national security purposes. As technology continues to advance, researchers will undoubtedly develop new and innovative ways to address this challenge and improve interception capabilities.

The main purpose is to pursue the target within the minimal control effort. The intercept problem involves a constant near the target velocity in the x-axis (V) and a slider velocity variable in the y-axis (v(t) axis).

The distance between the target and the pursuer at the initial time $t = 0$ is indicated by the term D, and the collision time is illustrated by $T = D/V$. The motion dynamics is defined as follows:

$$\dot{y} = v, \dot{v} = u \qquad (6.154)$$

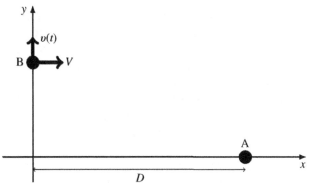

Figure 6.20 Two spacecraft – Target (a), and Follower (b).

where u describes the control variable, which shows the transverse acceleration. This problem can be expressed as an OCP as follows:

$$u_{min} = \arg \min_{u(t), t \in [0,T]} \int_0^T \delta \times u(t)^2 dt$$ (6.155)

$$\delta = [0.7, 1.5]$$

With the following boundaries:

$$\dot{y} = v, \dot{v} = u$$
$$y(0) = y_0, y(T) = 0, v(0) = 0$$ (6.156)

The above problem can be simplified as follows:

$$y_{min} = \arg \min_{u(t), t \in [0,T]} J(y)$$ (6.157)

where

$$J = \int_0^T d \times \ddot{y}(t)^2 dt$$ (6.158)

The reason for the simplification is that the initial and final conditions are defined on y. Generally, in classic methods, if the problem has uncertain parameters,

$$y(t) = \sum_{n=0}^N a_n T_n(t) = [T_0(t) T_1(t) ... T_N(t)][a_0(t) a_1(t) ... a_N(t)]^T$$ (6.159)

Consequently,

$$\begin{bmatrix} y(t_0) \\ \dot{y}(t_0) \\ \ddot{y}(t) \\ \vdots \end{bmatrix} = \begin{bmatrix} T_0(t) & T_1(t) & \cdots & T_N(t) \\ \dot{T}_0(t) & \dot{T}_1(t) & \cdots & \dot{T}_N(t) \\ \ddot{T}_0(t) & \ddot{T}_1(t) & \cdots & \ddot{T}_N(t) \\ \vdots & \vdots & \vdots & \vdots \end{bmatrix} \begin{bmatrix} a_0 \\ a_1 \\ a_2 \\ \vdots \end{bmatrix}$$ (6.160)

Solving the above classic problem is a simple task and there are different software tools for solving it, like @*Chebfun toolbox in Matlab*.

Now suppose the above model by considering its uncertainties; this case can be solved by the interval analysis methodology. Therefore, by using the Chebyshev inclusion direct method, the following results can be achieved.

Figure 6.21 displays the interval output of the state variable in the intercept problem using the Chebyshev inclusion direct method.

Figure 6.22 shows the interval output of the control variable in the intercept problem using the Chebyshev inclusion direct method.

Figure 6.23 shows the interval output of the control variable in the intercept problem using the Chebyshev inclusion direct method.

By applying the improved direct method in time intervals $[0, 1]$, it is observed that a relatively good error relative to the exact value is obtained. This error in both the upper and the lower bounds is about 1×10^{-4}.

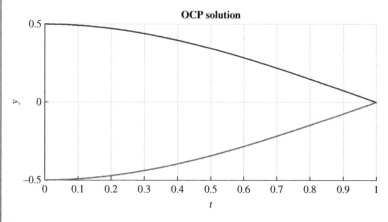

Figure 6.21 Interval output of the state variable in the intercept problem using the Chebyshev inclusion direct method.

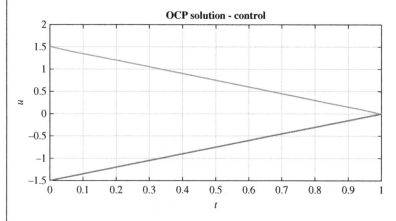

Figure 6.22 Interval output of the control variable in the intercept problem using the Chebyshev inclusion direct method.

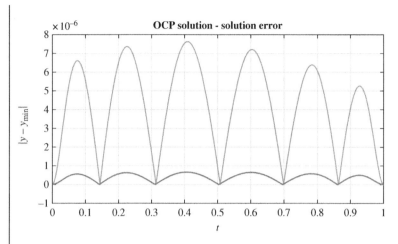

Figure 6.23 Interval output error of the intercept problem using the Chebyshev inclusion direct method.

Case Study 6.12   An Example of the Control Rule

The purpose is to find a control rule for the following function:

$$J = \int_0^1 \left(d_1 \times U(\tau)^2 + \delta_2 \times X(\tau)^2 \right) d\tau,$$

$$0 \le \tau \le 1$$

(6.161)

Such that, $\delta_1 = \delta_2 = [0.5, 1.5]$ and

$$U(\tau) = \dot{X}(\tau)$$

(6.162)

And the boundary conditions are

$$X(0) = \{0\}, X(1) = \left\{ \frac{1}{2} \right\}$$

(6.163)

The first thing that is required to use the Chebyshev method is to move the interval to $[-1, 1]$. After applying the transfer formula based on the previous sections, the required function will be obtained as follows:

$$J = \int_{-1}^1 \left(\hat{\delta}_1 \times u(t)^2 + \hat{\delta}_2 \times x(t)^2 \right) dt,$$

$$-1 \le t \le 1$$

(6.164)

Such that, $\hat{\delta}_1 = \hat{\delta}_2 = [0.25, 0.75]$ and

$$u(t) = 2\dot{x}(t) \qquad (6.165)$$

By considering the following boundary conditions:

$$x(-1) = \{0\}, x(1) = \left\{\frac{1}{2}\right\} \qquad (6.166)$$

The interval optimal control is based on a direct method with a multiobjective optimization algorithm (here we used NSGAII), the solution has been achieved as in Figure 6.24.

Figure 6.25 displays the state variable and the obtained interval obtained by the direct method and the central method with the interval final condition.

It can be observed that using the normal interval direct method here obtains a solution with a high overestimation (wrapping effect error). By applying the modified method and reducing the parameters of the system, the solution will be achieved as in Figure 6.26.

By analyzing the figure, we can observe that the implementation of the modified method, in this case, study has proven to be successful in enhancing

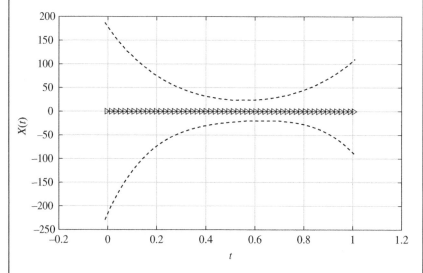

Figure 6.24 The state variable and the obtained interval obtained by the direct method (dashed line) and the central method (black-arrow sign).

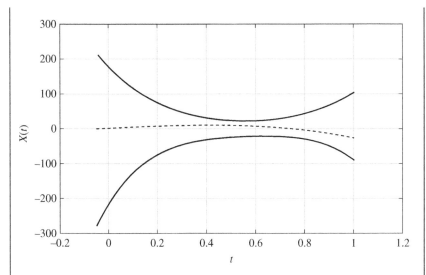

Figure 6.25 The state variable and the obtained interval obtained by the direct method (line) and the central method (dashed line) with the interval final condition.

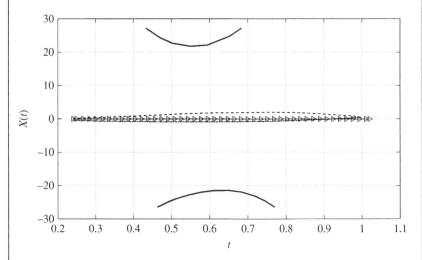

Figure 6.26 The state variable and the obtained interval obtained by the direct method (outer lines), modified direct method (dashed and dot-dashed lines), and the central method (arrow sign).

system performance. Taking a closer look at Figure 6.26, it becomes evident that the conventional interval Chebyshev method exhibits an interval range of [−300, 210], whereas the modified method significantly reduces this range to [−1, 1]. In simpler terms, it implies that without employing the modified method, the direct method would generate a substantial wrapping effect error at a very high level.

Case Study 6.13   The Interval Baseline Problem

The purpose of this problem is to determine the state and the control variables so that it can take a vehicle in the shortest time from one geographical location to another geographical location. This is an important problem for purposes such as guiding UAVs and preventing accidents.

By limiting the Baseline Problem in Transverse movement for the vehicle, the kinematics of the system can be achieved based on the Dubbin model.

$$
\begin{aligned}
\dot{x}_1(t) &= \delta_1 V(t) \cos(\theta(t)) \\
\dot{x}_2(t) &= \delta_1 V(t) \sin(\theta(t)) \\
\dot{V}(t) &= u_1(t) \\
\dot{\theta}(t) &= \delta_2 u_2(t)
\end{aligned}
\tag{6.167}
$$

where $p = (x, y)$ is the vehicle's position is in two directions, V is the vehicle speed, $\theta(t)$ is the angle of the movement, and $u = (u_1, u_2)$ describes the control variables, i.e., $V = \sqrt{\dot{x}^2 + \dot{y}^2}$.

In this equation, $z(t) = [x(t), y(t), \theta(t)] \in \mathbb{R}^3$ and $u(t) = [\dot{\theta}(t)] \in \mathbb{R}^1$. The speed of the vehicle is considered a constant value equal to 470 kts.

The OCP for this system has been achieved as follows:

$$
P_{\min} = \min_{u(t), t \in [0, t_f]} \int_0^{t_f} \delta \times J(p) dt
\tag{6.168}
$$

where $\delta = \delta_1 = \delta_2 = [0.7, 1.4]$ and t_f can be free, unspecified, or interval values. The boundary conditions are

$$
\begin{aligned}
P_0 &= (0,0), V_0 = 0.5, \psi_0 = \frac{\pi}{4} \\
P_f &= (1,1), V_f = 0.5, \psi_f = -\frac{\pi}{4}
\end{aligned}
\tag{6.169}
$$

These boundaries are used to restrict the vehicle's operational range. Therefore, with the help of this constraint, the prohibited areas can be induced arbitrarily.

The solution method for the problem will be the Dubins path (the shortest path between the two ends) if assuming a constant value for $V(x)$ and $u_1(t) = 0$. This problem can be solved by the direct method and by parameterizing the control variables based on Chebyshev inclusion functions with free coefficients. The time for moving in the Dubins path is considered 3.727 seconds. The system control track and the optimal path are shown in Figure 6.27.

Figure 6.28 shows the path response to the Dubins path.

It is evident from the results that when a random input is applied to the system, the solution falls within the designed interval. This example demonstrates a specific application of interval optimal control. Thus, when determining the interval boundaries in an optimal control system, it is crucial to only consider methods in which the values remain within the specified interval. Otherwise, selecting the wrong method can lead to inaccurate results.

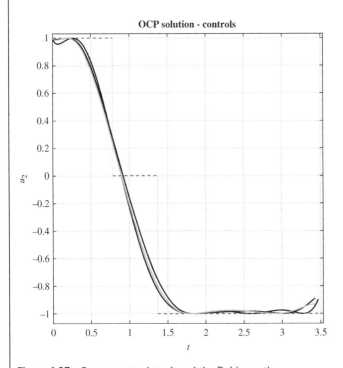

Figure 6.27 System control track and the Dubins path.

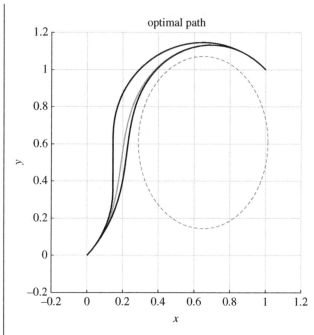

Figure 6.28 The path response to the Dubins path.

6.14 Conclusion

In this chapter, we have explored the necessary conditions for achieving optimal performance in interval systems and control systems. By employing the calculus of variations method, we have examined the essential conditions for optimizing a performance index in the presence of interval uncertainties, taking into account the system dynamics. Subsequently, we have applied these principles to analyze interval OCPs in various types of systems, including linear, nonlinear, and LQR systems. Building upon the findings from Chapter 3 and the methods proposed in previous chapters, we have presented a design for interval optimal control. This chapter covers both indirect and direct optimal control strategies. The direct method incorporates a modified approach that combines Chebyshev's orthogonal functions with multiobjective optimization to solve the studied system. To validate the proposed methods, we have conducted case studies and analyzed the obtained results. Through these analyses, we have demonstrated the effectiveness and applicability of the developed approaches.

References

1 Kirk, D.E., *Optimal control theory: an introduction*. 2004: Courier Corporation.

2 Naidu, D.S., *Optimal control systems*. 2002: CRC Press.

3 Wang, X., J. Liu, and H. Peng, *Symplectic pseudospectral methods for optimal control: theory and applications in path planning*. 2021: Springer Singapore, DOI: 10.1007/978-981-15-3438-6.

4 He, R. and H. Gonzalez. *Numerical synthesis of pontryagin optimal control minimizers using sampling-based methods*. in *2017 IEEE 56th Annual Conference on Decision and Control (CDC)*, 12–15 December 2017. 2017, Melbourne, VIC, Australia. IEEE.

5 Molloy, T.L., et al., *Inverse optimal control and inverse noncooperative dynamic game theory*. 2022: Springer.

6 Zadravec, D., et al., *Towards a comprehensive approach to optimal control of non-ideal binary batch distillation*. Optimization and Engineering, 2022. **23**: p. 1–31.

7 Simos, T.E., et al., *Unique non-negative definite solution of the time-varying algebraic Riccati equations with applications to stabilization of LTV systems*. Mathematics and Computers in Simulation, 2022. **202**: p. 164–180.

8 Guo, G. and R. Zhang, *Lyapunov redesign-based optimal consensus control for multi-agent systems with uncertain dynamics*. IEEE Transactions on Circuits and Systems II: Express Briefs, 2022. **69**(6): p. 2902–2906.

9 Din, A. and Q.T. Ain, *Stochastic optimal control analysis of a mathematical model: theory and application to non-singular kernels*. Fractal and Fractional, 2022. **6**(5): p. 279.

10 Chen, T., et al., *Modified evolved bat algorithm of fuzzy optimal control for complex nonlinear systems*. Romanian Journal Of Information Science And Technology, 2020. **23**: p. T28–T40.

11 Leal, U., et al., *Interval optimal control for uncertain problems*. Fuzzy Sets and Systems, 2021. **402**: p. 142–154.

12 Razmjooy, N. and M. Ramezani, *Optimal control of two-wheeled self-balancing robot with interval uncertainties using Chebyshev inclusion method*. Majlesi Journal of Electrical Engineering, 2018. **12**(1): p. 13–21.

13 Wang, C. and R.P. Agarwal, *Generalized hukuhara difference and division for interval and fuzzy arithmetic*, in *Dynamic Equations and Almost Periodic Fuzzy Functions on Time Scales*. 2022, Springer. p. 1–28.

14 Polak, E., *An historical survey of computational methods in optimal control*. SIAM Review, 1973. **15**(2): p. 553–584.

15 Burghes, D.N. and A. Graham, *Introduction to control theory, including optimal control*. 1980: John Wiley & Sons.

16 Bock, H., et al., *A direct multiple shooting method for real-time optimization of nonlinear DAE processes*. Nonlinear Model Predictive Control, 2000. **26**: p. 245–267.

17 Geng, F. and Z. Tang, *Piecewise shooting reproducing kernel method for linear singularly perturbed boundary value problems.* Applied Mathematics Letters, 2016. **62**: p. 1–8.

18 Wu, J., et al., *Uncertain analysis of vehicle handling using interval method.* International Journal of Vehicle Design, 2011. **56**(1–4): p. 81–105.

19 Akkouche, A., A. Maidi, and M. Aidene. *Solving optimal control problems by variational approach based on the Adomian's Decomposition Method.* in *Systems and Control (ICSC), 2013 3rd International Conference on.* 2013. IEEE.

20 Salahshour, S. and M. Khan, *Exact solutions of nonlinear interval Volterra integral equations.* International Journal of Industrial Mathematics, 2012. **4**(4): p. 375–388.

21 Berkani, S., F. Manseur, and A. Maidi, *Optimal control based on the variational iteration method.* Computers & Mathematics with Applications, 2012. **64**(4): p. 604–610.

22 Majumder, P., et al., *Application of Generalized Hukuhara derivative approach in an economic production quantity model with partial trade credit policy under fuzzy environment.* Operations Research Perspectives, 2016. **3**: p. 77–91.

23 Rodrigues, L., *Optimal control of a class of pseudo Euler-Lagrange systems.* Optimal Control Applications and Methods, 2017. **38**(2): p. 266–278.

24 Wu, J., et al., *Interval uncertain method for multibody mechanical systems using Chebyshev inclusion functions.* International Journal for Numerical Methods in Engineering, 2013. **95**(7): p. 608–630.

25 Heydari, A. and S.N. Balakrishnan, *Fixed-final-time optimal control of nonlinear systems with terminal constraints.* Neural Networks, 2013. **48**: p. 61–71.

26 Gouttefarde, M., D. Daney, and J.P. Merlet, *Interval-analysis-based determination of the wrench-feasible workspace of parallel cable-driven robots.* IEEE Transactions on Robotics, 2011. **27**(1): p. 1–13.

27 Ding, T., et al., *How affine arithmetic helps beat uncertainties in electrical systems.* IEEE Circuits and Systems Magazine, 2015. **15**(4): p. 70–79.

28 Zaks, A., et al., *Bitwidth reduction via symbolic interval analysis for software model checking.* IEEE Transactions on Computer-Aided Design of Integrated Circuits and Systems, 2008. **27**(8): p. 1513–1517.

29 Wu, J., *Uncertainty analysis and optimization by using the orthogonal polynomials,* Doctoral dissertation, University of technology, Sydney. Faculty of Engineering and Information Technology. 2015.

30 Huang, J.T., *Persistent excitation in a shunt DC motor under adaptive control.* Asian Journal of Control, 2007. **9**(1): p. 37–44.

31 Chu, H., et al., *Low-speed control for permanent-magnet DC torque motor using observer-based nonlinear triple-step controller.* IEEE Transactions on Industrial Electronics, 2017. **64**(4): p. 3286–3296.

32 Sola, H.B., et al., *Interval type-2 fuzzy sets are generalization of interval-valued fuzzy sets: toward a wider view on their relationship.* IEEE Transactions on Fuzzy Systems, 2015. **23**(5): p. 1876–1882.

33 Pappas, G.J., G. Lafferriere, and S. Sastry, *Hierarchically consistent control systems.* IEEE Transactions on Automatic Control, 2000. **45**(6): p. 1144–1160.

34 Patre, B.M. and P.J. Deore, *Robust state feedback for interval systems: an interval analysis approach.* Reliable Computing, 2010. **14**(1): p. 46–60.

35 Giusti, A. and M. Althoff. *Efficient computation of interval-arithmetic-based robust controllers for rigid robots.* in *Robotic Computing (IRC), IEEE International Conference on.* 2017. IEEE.

36 Fang, J., *The LQR controller design of two-wheeled self-balancing robot based on the particle swarm optimization algorithm.* Mathematical Problems in Engineering, 2014. **2014**, Article ID 729095, 6 pages.

37 Mansour, S.-M.B., J. Ghommam, and S.-M. Naceur, *Design and control of Two-Wheeled Inverted Pendulum Mobile Robot.* 13–15 September 2021, Zadar, Croatia. 2015.

38 Wuori, E. and J. Judy, *Rotational hysteresis for domain wall motion in the presence of demagnetizing fields.* IEEE Transactions on Magnetics, 1985. **21**(5): p. 1602–1603.

39 Shashikhin, V., *Robust stabilization of linear interval systems.* Journal of Applied Mathematics and Mechanics, 2002. **66**(3): p. 393–400.

40 Razmjooy, N., et al., *Comparison of lqr and pole placement design controllers for controlling the inverted pendulum.* Journal of World's Electrical Engineering and Technology, 2014. **2322**: p. 5114.

41 Gao, X., Y. Tian, and B. Sun, *Multi-objective optimization design of bidirectional flow passage components using RSM and NSGA II: A case study of inlet/outlet diffusion segment in pumped storage power station.* Renewable Energy, 2018. **115**: p. 999–1013.

42 Lewis, F.L., D. Vrabie, and V.L. Syrmos, *Optimal control.* 2012: John Wiley & Sons.

7

Conclusions

This book presents different methods for finding the confidence interval for optimal control problems in the presence of interval uncertainties. Based on the issues raised, these kinds of interval problems and their dynamics arise from different reasons due to measurement errors to the nature of the system itself.

As stated, there are several methods for solving problems with uncertain parameters, each of which has its drawbacks. Since the existence of various errors in natural systems is inevitably avoided and all uncertainties can be somehow contracted within a specific range, the use of interval analysis methods in solving these kinds of optimal control problems can be utilized as a valuable tool in solving these problems in a safe level for their robustness.

Many rules governing definite integers (including natural and complex) need to be established for the interval integers, which has caused many problems for researchers in dealing with these types of problems.

Moreover, there currently needs to be a specific method to use the interval analysis in optimal control problems. Therefore, in this book, we have filled out this vacancy by providing strategies and modifying them as much as possible. In general, the most important features of the book have been summarized as follows:

1) Modifying and developing interval methods based on central, forward, and backward representations.
2) Providing interval methods for solving the optimal control problems under interval uncertainties.
3) Using interval analysis to solve optimal control problems with direct and indirect methods.
4) Using a new interval numerical method based on Runge–Kutta models for solving dynamic problems with uncertainties.
5) Using the Chebyshev inclusion method to improve the time complexity and increase the confidence interval density.

Interval Analysis: Application in the Optimal Control Problems, First Edition. Navid Razmjooy.
© 2024 The Institute of Electrical and Electronics Engineers, Inc.
Published 2024 by John Wiley & Sons, Inc.

6) Using a piecewise version of the Chebyshev inclusion method to improve the time complexity and increase the indeterminate interval's density.

7) Extension of the proposed method for solving optimal control problems of linear and second-order closed loops.

8) Using multiobjective optimization algorithms as a direct optimal control method.

9) Using a recursive method for reducing the number of coefficients of the Chebyshev functions and speeding the compressibility of the confidence interval for solving optimal control problems by the direct approach.

10) Exploring and extending the stability of the interval system by linear matrix inequalities (LMIs).

Index

Interval Analysis: Application in the Optimal Control Problems, First Edition. Navid Razmjooy.
© 2024 The Institute of Electrical and Electronics Engineers, Inc.
Published 2024 by John Wiley & Sons, Inc.

Printed and bound by CPI Group (UK) Ltd, Croydon, CR0 4YY

16/04/2025

14658579-0001